P9-DMM-763

WorldShift 2012

Making Green Business, New Politics, and Higher Consciousness Work Together

Ervin Laszlo

Forewords by Deepak Chopra and Mikhail Gorbachev
Afterword by José Argüelles
With a special contribution by Tomoyo Nonaka

Inner Traditions
Rochester, Vermont • Toronto, Canada

Inner Traditions
One Park Street
Rochester, Vermont 05767
www.InnerTraditions.com

Copyright © 2009 by Ervin Laszlo

All rights reserved. No part of this book may be reproduced or utilized in
any form or by any means, electronic or mechanical, including photocopying,
recording, or by any information storage and retrieval system, without
permission in writing from the publisher.

Library of Congress Cataloging-in-Publication Data
Laszlo, Ervin, 1932–
  WorldShift 2012 : making green business, new politics, and higher
consciousness work together / Ervin Laszlo ; forewords by Deepak Chopra and
Mikhail Gorbachev ; afterword by José Argüelles ; with a special contribution
by Tomoyo Nonaka.
     p. cm.
  Includes bibliographical references and index.
  Summary: "A handbook for conscious change that could transform the
current world crisis into planetary renewal"—Provided by publisher.
  ISBN 978-1-59477-328-0
  1. Sustainability. 2. Change. 3. Civilization, Modern—1950– I. Title.
  HC79.E5L379 2009
  338.9'27—dc22

                                                                2009017087

Printed and bound in the United States by Lake Book Manufacturing

10 9 8 7 6 5 4 3 2 1

Text design and layout by Jon Desautels
This book was typeset in Sabon with Futura used as the display typeface

To send correspondence to the author of this book, mail a first-class letter
to the author c/o Inner Traditions • Bear & Company, One Park Street,
Rochester, VT 05767, and we will forward the communication.

WORLDSHIFT 2012 IS THE CLUB OF BUDAPEST'S HANDBOOK OF CONSCIOUS CHANGE

Founded by Ervin Laszlo in 1993, the Club of Budapest (www.clubof budapest.org) is an informal international association of individuals of high ethical standard and moral integrity dedicated to developing new ways of thinking and acting to help resolve the social, political, economic, and ecological challenges of the twenty-first century. With its eminent members and dedicated national clubs active in many parts of the world, the club initiates and sustains a dialogue between different belief systems and worldviews in order to cocreate effective strategies for responsible action with a global focus (see www.worldshift2012.org).

HONORARY MEMBERS

Oscar ARIAS
statesman, Nobel Peace
  Laureate

A. T. ARIYARATNE
Buddhist spiritual leader

Jose ARGÜELLES
Anthropologist/spritual leader

Thomas BERRY
theologian/scientist

Karlheinz BÖHM
actor/activist

Deepak CHOPRA
physician, spiritual leader

Paulo COELHO
writer

Mihaly CSIKSZENTMIHALYI
psychologist

The XIVth DALAI LAMA
spiritual leader/Nobel Peace
  Laureate

Riane EISLER
feminist historian/activist

Vigdis FINNBOGADOTTIR
political leader/Nobel Peace
  Laureate

Milos FORMAN
film director

Peter GABRIEL
musician

Hans-Dietrich GENSCHER
statesman

Jane GOODALL
scientist

HONORARY MEMBERS (CONT.)

Rivka GOLANI
musician

Mikhail GORBACHEV
opinion leader/statesman

Arpád GÖNCZ
writer/statesman

Václav HAVEL
writer/statesman

Hazel HENDERSON
economist/activist

Bianca JAGGER
activist

Miklós JANCSÓ
film director

Ken-Ichiro KOBAYASHI
orchestra director

Gidon KREMER
musician

Hans KÜNG
Christian spiritual leader

Shu-hsien LIU
Chinese philosopher

Eva MARTON
opera singer

Zubin MEHTA
orchestra director

Edgar MITCHELL
scientist/astronaut

Edgar MORIN
philosopher/sociologist

Robert MULLER
educator/activist

Gillo PONTECORVO
film director

RAFFI (Raffi Cavoukian)
singer, children's cause activist

Mary ROBINSON
political and human rights leader

Peter RUSSELL
philosopher/futurist

Masami SAIONJI
Japanese spiritual leader

Karan SINGH
Hindu spiritual leader

Rita SÜSSMUTH
political leader/activist

Sir Sigmund STERNBERG
interfaith spiritual leader

Desmond TUTU
spiritual leader /Nobel Peace
Laureate

Liv ULLMANN
actor/director

Mohammad YUNUS
economist, Nobel Peace
Laureate

Contents

WorldShift: *A worldwide* shift *from a path of unsustainability, conflict, and confrontation to a path toward sustainability, well-being, and peace.*

Foreword

Deepak Chopra

We are already living in two worlds. One world moves ahead by inertia from the past, like a massive luxury liner drifting at sea, while the other steps into the unknown, like a child entering the woods for the first time. On the front pages of newspapers and on the evening news, the first world gains the lion's share of attention. A new crisis deepens yesterday's crisis in Africa or the Middle East. A fresh humanitarian outrage taints a faraway society. One war replaces another. Despite the sameness of these events, they constitute the news of the world as far as the mass media shows it. Yet this world of inertia and non-change is deceptive. Beyond crisis-driven news, another world is rising. In this book Ervin Laszlo maps this new world and its promise—no one I know is more acutely attuned to it.

The first world is a solid wall that looks impregnable, yet behind it people no longer feel protected. They dream of a shift in consciousness, the revolution that needs only to be asked for and it will begin. Material events are but the outward display of consciousness. Paying attention only to the world of inertia and non-change is like dwelling in illusions. In the 1980s the annual May Day marches of massive Soviet armaments through Red Square didn't reveal that the Communist system was about to collapse. Armies, wars, ecological disaster, unbridled greed and corruption, skyscraper cities springing up like weeds, a deluge of pesticides and pollutants, streams of refugees without a homeland, tyranny spreading violence without check, pandemic disease: these are the fruits of consciousness, too, but a kind of consciousness that is stuck and unable to raise itself above its self-created problems.

What is so admirable about WorldShift is that we are led directly to this diagnosis and then offered a solution that is also in consciousness. Ervin Laszlo joins a small group of deeply versed thinkers who grasp that all experience occurs in consciousness and nowhere else. This insight saves him from the trap of replacing one style of materialism with another. Those on the right who try to impose democracy on societies that have no basis for democracy in their traditions are hitting their heads against an invisible fact: what you aren't aware of, you can't change. Those on the left who promote globalism as new partnerships in trade and economics miss the point that the Americanization of the world—bringing rampant consumerism to traditional societies—can turn out to be destructive, not just to valuable folkways but to the planet. Only a shift in consciousness can save the world from impending perils that are also rooted in consciousness.

Fortunately, the second world—the world of timely change—is poised to save the first. People are beginning to confront the world-shaping trends that this book points out so succinctly:

- ► The dispossessed of the earth are rising and won't be suppressed in their quest for prosperity.
- ► The planet's sustainable resources will be depleted unless human beings find a sustainable way of life in accord with Nature.
- ► Materialism has reached its historical apogee and will decline or self-destruct through accelerating degradation of the ecology.

As viewed from the first world, these are such overwhelming threats that the response of governments has been to look the other way or to make little more than symbolic gestures at reform. From the perspective of the second world, it's no surprise that governments are stymied, because the policies that despoiled the earth can't be expected to renew it, either by doing less or doing more.

Among his many utterly cogent points, Ervin Laszlo declares that we need a new way to be happy. For me, this is the deepest and most salient point. When an American housewife drives her car to the supermarket, purchases brightly packaged processed food, leaves a full garbage can out on the curb, and sprays a can of insecticide to kill the aphids in her rose garden, none of these actions seem destructive—she's simply doing the ordinary things we all do in our pursuit of happiness. But happiness based on waste, toxins, depletion of fossil fuels, and endless consumer goods—the paradise we have all chased since the end of World War II—can't be sustained. Still less can we sustain the massive military forces that serve to shut out 90 percent of humankind so that the privileged 10 percent can promote a worldview that over time will spell the end of their existence along with everyone else's.

When stated like that, the future seems dire. So it comes as a relief that this handbook for conscious change goes beyond superficial pessimism or optimism, offering instead a new way to be happy. Without a doubt the outmoded world of materialism is leading to greater unhappiness, through pollution, overpopulation, lack of nourishing food and water, and the loss of natural habitats: a sizable percentage of the world's population already experiences these deficits. Timely change through a shift in consciousness can bring about a new model of happiness based on the principles of higher consciousness. The Indian spiritual teacher J. Krishnamurti, not one for blithe cheerfulness, used to say, with justice, that to live beyond the simple needs of a modest house, good food, a few well-fitted clothes, and other necessities was not to become more happy. Quite the opposite. To the extent that we look outward for happiness, the final result will be boredom, stagnation, and bitter disappointment.

WorldShift is about an outer world built on inner realization. Such a world is possible, as this book shows, and indeed is already being born in the hearts of millions of people. The front pages of newspapers and the evening news on CNN aren't reporting this shift, but scientist-philosophers like Ervin Laszlo have already laid a foundation upon which magnificent edifices may soon rise. Higher consciousness, which has been evolving through human beings for centuries, is seeding a garden of hope and promise.

Foreword

Mikhail Gorbachev

Dear Reader: This handbook for the twenty-first century speaks to you in person. Indeed, this is a message that is addressed to you and to all of us. It is written in the hope that you will not only read it, but will also think through the things it tells you. And with the further hope that you will draw the necessary conclusions, your own conclusions, for yourself, your family, your friends, and everyone close to you.

Why did the author of this book, Ervin Laszlo—the famous scientist and humanist, and president of the Club of Budapest—choose this specific form, the form of a message addressed to us, to each and every one of us, his readers?

In general, when someone's future or some aspect of the world that surrounds us in daily life is in question, we see the gist of the issue fairly easily and quickly. We see the advantages and dangers and draw the conclusions, deciding what steps to take. This is natural, it corresponds with our habits, it is part of our everyday thinking and behavior. For someone living in a complex world, common sense dictates that he or she must think about whether or not to adapt to the given circumstances, or try to change them. The situation is different when we confront problems that affect the whole of the world, the destiny of all humankind. We are not used to questions of such dimensions. It may seem that they are far away and that some time, somehow, they will be solved, indeed that someone "up there" is taking care of them. "Why us? We are only little people."

This is why the book in our hand, dedicated to global, world-encompassing problems, addresses us in plain and logical language and marshals persuasive evidence. This makes our task easier. The task is simple. Get down to the basics, understand that global problems are not foreign to us. They are our problems. We are all touched by them, and touched by them not any less than we are by ordinary, everyday things. And it is we, each one of us, who not only can understand these problems, but can also do something significant to overcome them.

WHAT IS AT STAKE?

The fact is that with the passing of time a whole pyramid of diverse problems has been accumulating in every part of the world: social, political, economic, and cultural problems. Contradictions have appeared in society—in a different way in each country, but present all the same—and they have created conflicts and crises. Even wars. The relationship between humans and nature has become more and more complex and strained. The air has become poisoned, rivers polluted, forests decimated. The numbers of contradictions keep growing, and they are becoming deeper. Symptoms of illness in society became obvious.

In our various ways all of us, in every part of the world, have expressed our dissatisfaction with this state of affairs, have demanded changes, and are still demanding them. Isn't this story familiar? I think it is.

People everywhere began to show discontent with this state of affairs and demand changes. Violent movements have arisen, such as strikes and disturbances. Society has entered a period of crisis. How will this crisis be resolved? This is difficult to predict. Society's sickness affects every single citizen and threatens everyone with suffering. The end result may be an explosion, a bloodbath that nobody wants, yet that comes about spontaneously.

ANOTHER WAY OUT

Is there another way out, a path beyond the crisis? The book in our hand gives an answer: yes, there is another way. We must not wait until society's crisis reaches the danger point. We must act! Every person can act. If everyone does his or her bit, together we can accomplish what is necessary. We can make an impact on those who decide the politics and the destiny of society, and motivate them to begin making the necessary changes. Changes that will not only resolve the crisis, but also show us a path of survival, of healthy development for people and nature, and a better quality of life for all. That is our salvation.

The human community has reached the point where it is obvious that events cannot be allowed to take their own course. It is necessary to make a turn that would change the character and the content of development for the benefit of humankind. We have already become conscious that change is truly necessary. Now we must understand what exactly we must do to avoid the worst, and how we must do it. This book will help us to evaluate the current situation of our planet and to find the path we must take.

Preface

The world we have created is not sustainable. Finding ways to mitigate the worst effects of today's economic and financial crisis doesn't mean that we have reached the end of the crisis, or even the beginning of its end; we have only reached the end of its beginning. For other crises will follow—ecological and demographic crises, as well as those triggered by the squeezing of energy, water, and mineral resources—crises that will embrace all peoples and states, and all segments of society. Whether we realize it or not, we have entered a state of global emergency.

A condition of global emergency is not a cause for pessimism and despair; it's a call for action. A crisis, whether local or global, allows many things, including some highly constructive things. There is just one thing that it doesn't allow: going on as before. For then the crisis would turn into catastrophe.

Crisis calls for change; not cosmetic change, not piecemeal change or short-term remedial change. It calls for fundamental, system-wide change, for *conscious change that comes in time*. Such a change is now possible, for crisis weakens the hold of the old and permits the rise of the new. In the fall of 2008 the financial crisis has driven constructive change in U.S. politics, and further crises can drive positive change in America, and in other parts of the world.

As *Wei-Ji*, the Chinese word for "crisis," indicates, crisis is danger as well as opportunity. We must be aware of both.

The danger is not negligible. Crisis unsettles and threatens the old and the established. Super-rich financial institutions go bankrupt from one day to the next; wealthy states fight a losing battle with ever-greater burdens of debt; the entire financial system threatens to collapse. New economic superpowers enter the scene, and their insistence on conventional forms of growth could exceed the resources of the planet and its capacity to absorb waste.

Earth itself is transforming under our feet. Changing patterns of rain, violent storms, and prolonged periods of drought threaten vast tracts of land, the breadbaskets of hundreds of millions. The air is heating up. On New Year's Eve 2007 Russians celebrated in Red Square without a trace of ice and snow; in January 2008 New Yorkers walked in Central Park in shirtsleeves; the center of Greenland is taken up by an unfrozen lake the size of Lake Michigan, Lake Superior, and Lake Erie combined, and there is hardly any of the legendary snow left on top of Kilimanjaro. Giant masses of sheet-ice slide from the Antarctic shelf into the ocean and raise sea levels worldwide, threatening the habitat of hundreds of millions and flooding vast tracts of productive land.

The climate is just one of the many dangers facing us; together with financial crashes and economic fluctuations, it's the most visible. Besides the climate and the economy, there are hosts of other "unsustainabilities": poverty, inhuman living conditions, hunger, crime, terrorism, war, intolerance, and the revolt and frustration they entail. These are less evident but just as dangerous. If we don't begin effective and timely change, spreading chaos will escalate into catastrophe. It would lead to the end of today's world, and could terminate the tenure of humankind on the planet. We either change in time, or suffer the consequences.

But danger is only one face of crisis: opportunity is another. Now that crisis has come, we must seize the opportunity it brings—the opportunity to change consciously, and in time.

Conscious change is our opportunity. A stable system, whether it's a society or an ecology, doesn't change readily; it's committed to the status quo. An unsustainable system is more prone to change, because it *needs* to change. It could change fast. The question is, how does it change? Haphazardly, under critical stress, or consciously and effectively, in time to reach more sustainable, less crisis-prone conditions?

Conscious change could bring the change we need, and bring it in time. We have all the requisites for it. There are enormous and as yet unexploited resources at our disposal: human, financial, and natural. An entire spectrum of new energy and resource technologies is coming on line, and more informed and responsible leaders are taking the helm in politics and in business. The need for change is penetrating the thinking of ever more people. New insights into our connections to each other and to nature are emerging at the cutting edge of the sciences, and the new cultures of peace, sustainability, and responsibility are beginning to live them. Individually and collectively, as conscious humans and as a conscious species, we are waking up. We need not become extinct. We can change.

This is a handbook of conscious change—change that shifts today's world from the path of deepening crisis to a kind of development that can bring peace and well-being to the human family. This would be an epochal shift, a *WorldShift*. Creating it is feasible. But whether it will be actually created is not decided yet: it depends on what we do today. And that, in turn, depends on our values, our ethics, and our consciousness.

Part One

THE DESTINY CHOICE

We live in a curious world, and have a curious attitude toward the world. The environment is degenerating, the climate is changing, but most of us are preoccupied with making money and maintaining our privileges—we keep juggling for position on the deck of the *Titanic*.

> There is an overproduction of food in many parts of the world, but more than a thousand million go hungry; every six seconds someone dies of starvation.
>
> Six million children a year die for lack of food, while 155 million are overweight.
>
> The wealth stockpiled by a few hundred billionaires is more than the income of three billion people, nearly half of humanity.
>
> We face grave problems and a gamut of unresolved tasks, but put millions out of work.
>
> The problems we face call for long-term solutions, yet our criterion of success is the bottom line in annual or semi-annual corporate profit and loss statements.
>
> We stockpile hi-tech weapons that are more dangerous than the conflicts they are meant to overcome.

We fight cultural intolerance and religious fundamentalism, but have been, and some of us still are, willing to subscribe to virulent forms of nationalism under the banner of patriotism and national security.

Forty minutes of the solar radiation that reaches Earth, fully used, would cover all of humanity's energy needs for a year, but we insist on squeezing the last drop of polluting and already peaked oil from under the deserts and the seas.

We have millions of intelligent and responsible women ready to play a responsible role in society, but we don't give them a fair chance in business, in education, and in civic life.

We tell children to abide by the golden rule "treat others as you expect others to treat you," but we don't treat other people, other states, and other businesses as we expect other people, states, and businesses to treat us.

We bring vast herds of sensitive animals into the world for the sole purpose of slaughtering them, a practice that has highly questionable ethical and even health implications and is taking an enormous amount of water and grain.

This list could go on and fill a volume of its own. The fact is that our world is neither rational nor sustainable. It's unstable and unfair, and prone to violence. It could get more unstable, unfair, and violent before long. We had better wake up and begin to put our house in order.

ONE

What Would Happen If . . .

Just what *would* happen if we let things go on as they are? And what *could* happen if we wake up and change?

THE PATH TO BREAKDOWN

Here, first, is the likely BAU (business-as-usual) scenario.

- There is no real change in the way business is conducted in the world, resources are exploited, and energy is produced. This leads on the one hand to worsening economic gaps and conflicts, and on the other to accelerating global warming and climate change.
- In some regions the warming of the atmosphere produces drought, in others devastating storms, and in many areas it leads to harvest failures. As the Antarctic glaciers melt, giant masses of ice slide from the land-shelf into the ocean, flooding cities, towns, and villages and productive lands. Hundreds of millions are made homeless and face starvation.
- Massive waves of destitute migrants flow from coastal regions and inland areas afflicted with lack of food and water toward regions where basic resources remain accessible. The migrants overload the human and natural resources of the receiving countries and create conflict with the local populations. International

3

relief efforts provide emergency supplies for thousands, but are helpless when confronted with millions.

- In futile attempts to stem the tidal wave of destitute people, India completes its wall along the border with Bangladesh, the U.S. along the Mexican border, and both Italy and Spain build walls to protect their northern regions from the overrun southern areas.

- The world's population fragments into desperate masses facing imminent famine and disease, and states and economies just as desperate to protect themselves. The conflicts create unsustainable stresses and strains in international relations: social and economic integration groups and political alliances break apart. Relations collapse between the U.S. and its southern neighbors, the European Union and the Mediterranean countries, and India and China and the hard-hit Southeast Asian states.

- Global military spending rises exponentially as governments attempt to protect their territories and reestablish a level of order. Strong-arm régimes come to power in the traditional hot spots as food and water wars erupt between states and populations pressed to the edge of survival.

- Terrorist groups, nuclear proliferators, narco-traffickers, and organized crime syndicates form alliances with unscrupulous entrepreneurs to provide arms, drugs, and essential goods at exorbitant prices. Governments target the terrorists and attack the countries suspected of harboring them, but more terrorists take the place of those they imprison or kill.

- Hawks and armaments lobbies press for the use of powerful weapons to defend their territories and interests. Regional wars fought initially with conventional arms escalate into wars conducted with weapons of mass destruction.

- The environment, its productive processes and vital heat balance impaired, is no longer capable of providing food and water for more than a fraction of the surviving populations. The world's critically destabilized economic, social, and political system collapses. Chaos and violence engulf peoples and states on the five continents.

THE PATH TOWARD BREAKTHROUGH

BAU is not the only scenario available to us; we could also change. Here is a possible timely change scenario.

- The threat of terrorism and war, together with rising poverty and growing environmental degeneration, trigger the widespread recognition that the time to change has come. In country after country, an initially small but soon rapidly growing nucleus of people pull together to confront the dangers of the crisis and seize the opportunity it offers for change.
- The rise of popular movements for sustainability and peace leads to the election of political leaders who support policies and projects of economic cooperation and social solidarity. Forward-looking states monitor the dangerous trends and provide financing for the urgently needed economic, ecological, and humanitarian initiatives.
- Non-governmental organizations link up to revitalize regions ravaged by environmental degradation. Emergency funds are provided for countries and regions afflicted by drought, violent storms, coastal flooding, and harvest failures.
- Military budgets are reduced and in some states eliminated, and the resulting "peace dividends" are assigned to increase the production of staple foods, safe water, basic supplies of energy, and essential sanitation and health services for the needy populations.
- Country after country shifts from fossil-fuel-based energy production to alternative fuels, reducing the release of greenhouse gases into the atmosphere and slowing global warming. A widely networked renewable energy system comes on line, contributing to food production, providing energy for desalinizing and filtering sea

water, and helping to lift marginalized populations from the vicious cycles of poverty.

- Leading business companies replace the classical strategy of seeking to maximize only their own profit and growth with concern with social and ecological responsibility. On the initiative of enlightened managers a voluntarily self-regulating social market economy is put in place, and it is given full support by the newly elected socially and ecologically aware political leaders.

- As the new energy system and the self-regulating social market economy begins to function, the balances of nature are no longer violently impaired and the biosphere begins to heal itself.

- Access to economic activity and technical and financial resources becomes available to all countries and economies. Frustration, resentment, animosity, and distrust give way to a spirit of cooperation, liberating the spirit and enhancing the creativity of a new generation of locally active and globally thinking citizens. Humanity is on the way to a peaceful and sustainable, diverse yet cooperative civilization.

TWO

Why We Need Timely Change
THE UNSUSTAINABILITY OF
TODAY'S WORLD

The reason why we need to change should now be clear: the world we have created is structurally and chronically, but not incurably, unsustainable. But what is the nature, and what are the roots, of this unsustainability?

Understanding what unsustainability means is not difficult: it means that a process, or a condition, cannot continue without change. Either it changes or it breaks down. This situation on our planet is like the predicament of the sorcerer's apprentice: we have created conditions that we can no longer control. Or, like Aladdin, we have liberated the genie from the bottle and now it runs our lives. And if we don't change in time, it will *ruin* our lives.

We need to take control. This means change, but not superficial or haphazard change. It means conscious change, addressing the very conditions that make for unsustainability.

This calls for tackling the more difficult question: understanding the real causes of unsustainability. There is a great danger of oversimplification here. In most cases only one of the causes of unsustainability is taken into account, and the entire condition of unsustainability is ascribed to it. For example, it is often said that the root cause is the climate. Global unsustainability is due to global warming and its various effects. Or the cause is poverty. The breakdown of societies is due to the deepening and spreading of the centers of poverty and this creates famine and misery, generates frustration and hate, gives rise to maverick violence and terrorism and leads to the organized violence of war. Or the shortage of water, or the diminishing of productive soils, or urban crowding, or unemployment, or the price of energy . . . there are myriad causes we could name.

But the truth is that for unsustainability there is not just one basic cause, and to think otherwise is more than just to make a mistake; it's to make a dangerously misleading mistake. For then we concentrate on that cause at the expense of the others.

Unsustainability is a systemic process and needs to be systemically understood. When the integrity of a complex system such as a human society or the environment (which is also a complex system) is impaired, single-factor remedies are bound to fail. As in holistic medicine, the condition of the "patient" needs to be taken into account in its totality. This is not an easy task, but it's by no means impossible. Carrying it out starts with looking at the systemic roots of the unsustainability that besets today's world.

A detailed enumeration of the strands of unsustainability in the world would require a good-sized encyclopedia, but we can proceed in a more manageable way. We can look at systemic clusters of unsustainabilities, and see how the elements that constitute a cluster hang together, and how each cluster interacts with the others.

We start with the cluster of unsustainability in the ecology, and then move to the clusters of unsustainability in the economy and in society.

UNSUSTAINABILITY IN THE ECOLOGY

The way we exploit the environment is intrinsically unsustainable. According to former UN Assistant Secretary-General Robert Muller, each minute 52 acres of tropical forest are lost, 50 tons of fertile topsoil are blown off, and 12,000 tons of carbon dioxide are added to the atmosphere. Each hour 1,693 acres of productive dry land become desert, and each day 250,000 tons of sulfuric acid fall as acid rain in the Northern Hemisphere. An estimated 100,000 chemical compounds are injected into the land, rivers, and seas; millions of tons of sludge and solid waste are dumped into the oceans; and billions of tons of CO_2 are released into the air.

Water. Water is a major element in the cluster of ecological unsustainabilities. The amount of available fresh water is diminishing rapidly; over half the world's population faces water shortages. On average, 6,000 children are dying each day of diarrhea caused by polluted water.

In the past, the planet's available freshwater reserves were adequate to satisfy human needs: in 1950 there was a potential reserve of nearly 17,000 m^3 of fresh water for every person then living. Since then the rate of water withdrawal has been more than double the rate of population growth, and in consequence in 1999 the per capita world water reserves decreased to 7,300 m^3. If current trends were to continue, in the year 2025 there would be only 4,800 m^3 of reserves per person. This amount, combined with uneven access to the reserves, would create major health hazards in many parts of the world.

Already today, about one-third of the world's population doesn't have access to sufficient supplies of safe water, and by 2025 two-thirds of the population will live under conditions of critical water scarcity. Whereas Europe and the United States will have half the per capita water reserves they had in 1950, Asia and Latin America will have only a quarter. The worst hit countries will be in Africa, the Middle East, and south and central Asia, where the available supplies may drop to less than 1,700 m^3 per person.

Productive land. The progressive reduction of the amount of land capable of producing food is another threatening trend. The Food and Agriculture Organization (FAO) estimates that on the global level there are 7,490 million acres of high-quality cropland available, 71 percent of it in the developing world. This quantity is decreasing due to soil erosion, destructuring, compaction, impoverishment, excessive desiccation, accumulation of toxic salts, leaching of nutritious elements, and inorganic and organic pollution owing to urban and industrial wastes.

In some parts of the world, this trend augurs major food shortages. China has a population that is five times that of the United States, but has only one-tenth as much cultivated land; it is feeding 24 percent of the world's population on 7 percent of the world's agricultural land. This small percentage is further diminishing. Due to urban sprawl and the construction of roads and factories, 37 million acres of China's cultivated land have already been converted to nonagricultural use. Of the remaining 247 million acres one-tenth is highly polluted, one-third is suffering from water loss and soil erosion, one-fifteenth is salinized, and nearly 4 percent is becoming arid and turning into a desert.

Worldwide, 12 to 17 million acres of cropland are lost per year. If this process continues, some 741 million acres will be lost by mid-century, leaving 6.67 billion acres to support 8 to 9 billion people. This would be catastrophic, as the remaining 0.74 acres of productive land per person could only produce a bare subsistence level of food. The prospects are actually still worse, since giant masses of ice are sliding from the Antarctic land-shelf into the oceans, threatening a rapid rise in sea levels. A major rise would flood large tracts of land in coastal areas, depriving hundreds of millions of habitat, as well as of food and water.

Air. The way we pollute the atmosphere is also unsustainable. The amount of air that humans, and even all organisms taken together, need is minuscule compared to the size of the atmosphere that surrounds the planet. But here, too, it's a question of quality rather than quantity. Polluted air and air of inadequate oxygen content is of little use. Yet the oxygen content of the atmosphere is diminishing, and its carbon dioxide and other greenhouse gas content is increasing.

Since the middle of the nineteenth century oxygen has decreased mainly due to the burning of coal; it now dips to 19 percent of total volume over impacted areas and 12 to 17 percent over major cities.

At 6 or 7 percent of total volume, life can no longer be sustained. At the same time the share of greenhouse gases is growing. Two hundred years of burning fossil fuels and cutting down large tracts of forest have increased the atmosphere's carbon dioxide content from about 280 parts per million to over 350 parts per million, and its CO_2 content is increasing more and more rapidly.

The influx of gases due to industrial sources is matched by the growing influx from nature, indirectly and unwittingly triggered by human activity. In Siberia, an area of permafrost spanning a million square kilometers—the size of France and Germany combined—started to melt for the first time since it formed at the end of the last ice age 11,000 years ago. Russian researchers found that what was until recently a barren expanse of frozen peat is turning into a broken landscape of mud and lakes, more than a thousand kilometers across. The area, the world's largest peat bog, has been producing methane since it formed at the end of the last ice age, but most of the gas has been trapped under the permafrost. It is now being released.

Global warming and climate change. Changes in the chemical composition of the atmosphere create the hothouse effect. A shield in the upper atmosphere prevents heat generated at the surface from escaping into surrounding space. Global warming is an indisputable fact: in recent years the average global temperature has risen significantly, and the warming trend is accelerating. And climate models show that even minor changes in the atmosphere can produce major effects, including widespread harvest failures, water shortages, increased spread of diseases, the rise of the sea level, and the die-out of large tracts of forest.

Unsettled weather patterns due to warmer air masses have already caused major damage. In 2004 violent storms in the Philippines killed 669 people and another 695 were reported missing. The following year Hurricane Katrina killed more than 1,800 people in and around New Orleans and caused damages estimated at over $81 billion. In 2007 floods in India, Bangladesh, Nepal, and Bhutan were reported the worst in history, with an abnormal volume of monsoon rain killing more than

2,000 people and leaving 30 million homeless. In the same year heat waves triggered forest fires in Greece that killed over 80 people and caused 3,418 deaths throughout Europe. In the winter of 2009 near-hurricane force winds repeatedly battered the coast of France, and an unprecedented firestorm waged devastation in southeast Australia. The number and intensity of cyclones over the British Isles and the North Sea are projected to increase in the coming years, producing losses of life and agricultural productivity throughout Europe.

The causes of global warming are still debated. Is it due to human activity, or to natural causes? There were other warming periods in the history of the earth; geologists speak of alternating hot and cold stable states—"hothouses" and "ice-houses." The best known previous hothouse occurred 55 million years ago, when between one and two terratons of carbon dioxide were released into the air, most likely by the impact of a large meteorite. This was associated with an 8-degree rise in temperature in the Arctic zones and a 5-degree rise in the tropics. It took about 200,000 years for the Earth's atmosphere to return to its previous level.

Some people claim that today's warming is mainly due to natural causes, at the most exacerbated by human activity. A new cycle in the fusion-processes that generate heat in the Sun sends more solar radiation to Earth and heats up the atmosphere. Unfortunately, the same people who ascribe global warming to solar activity for the most part also dismiss the need for doing something about it—after all, what can we do to change the chemistry of the Sun? This, however, is a grievous mistake. We can't do anything about the chemistry of the Sun, but we *can* do something about reducing its effect on Earth. Doing so is necessary, whether the warming we experience is due to the Sun or has an anthropic component. The fact remains that global warming is producing climate change and ecological stress, and reducing the chances of well-being throughout the planet.

Today, CO_2, methane, and other hothouse gases are at historically unprecedented levels. We had injected a full terraton of carbon dioxide into the atmosphere in the twentieth century alone, and are presently injecting another terraton in less than two decades. The problem is not just the *amount* of CO_2 released into the atmosphere, but the *rate* at which it is released. In the hot-house period 55 million years ago the CO_2 correlated with global warming was released over a period of 10,000 years. Now we release a similar amount—estimated at one terraton—in a few decades.

The rapid injection of carbon dioxide makes it impossible for Earth's ecosystems to adjust. In the oceans the explosive growth of CO_2 at the surface makes the water too acid for the survival of shell-forming organisms, the marine species that is the basis of the chain of life in the seas. Seawater acidity has already increased by 30 percent and is likely to treble by the end of the century. On land, the absorption of carbon dioxide is reduced by the destruction of the ecosystems that had historically sustained a stable climate. Now as much as 40 percent of these systems are gone, due to acid rain, urban sprawl, and the injection of a variety of toxins into the soil.

A number of processes, once triggered, feed on themselves. Due to warmer air, ice is melting at the poles. Measurements indicate that temperatures in the western Arctic are at a 400-year high. If this trend continues, the Arctic Ocean will be ice free well before the end of the century; summer ice could disappear as soon as 2013. This will accelerate global warming, since the surface of liquid water is darker than ice and absorbs more heat—and this in turn melts more ice. The process also takes place at the South Pole. Because ice-melt is faster than new snowfall, in more and more spots the continent's ice-cover is changing color and is absorbing more heat. Entire glaciers are sliding off the land-shelf and dump an estimated 103 billion tons of ice a year into the water, raising sea levels the world over.

The methane released by the West-Siberian peat bog is a similar feedback process. The melting of the permafrost changes the ice-cover into water and speeds up the melting—which in turn increases the release of methane. The peat bog may hold as much as 70 billion tons of this gas, a quarter of all of the methane stored in the ground around the world.

Calculations show that it could release around 700 million tons of CO_2 into the atmosphere each year, about the same amount released from all of the world's wetlands and from agriculture. This would double atmospheric levels of the gas, leading to a ten to twenty-five percent increase in global warming.

Prolonged drought is the element of climate change that poses the greatest threat to human survival, for it limits global food production. This is a worldwide problem. California is facing the worst drought in recorded history; thousands of acres of row crops have already been fallowed. The average dimension of the snow pack in the northern Sierra, with some of the state's most important reservoirs, is only 49 percent of normal. In Texas the drought is reaching historic proportions. Near Austin and San Antonio the lack of rain has been exceeded only once before, in 1917–18. It's estimated that 88 percent of Texas is experiencing abnormally dry conditions, and 18 percent extreme or exceptional drought.

The worst drought in half a century has turned Argentina's once-fertile soil to dust and created a state of emergency. The country's wheat yield for 2009 is expected to be 8.7 million metric tons, compared to 16.3 million in 2008. Brazil, the world's second-biggest exporter of soybeans and third-largest exporter of corn, has cut its outlook for these crops after assessing desiccation damage to plants in the drought-stricken regions.

In Northern China the drought has been the worst in 50 years. It's creating water shortage for 4.37 million people and 2.1 million live-stock. The Chinese government has resorted to cloud seeding, and allocated 86.7 billion yuan (about $12.69 billion) to the drought-hit areas. Australia, in turn, has been experiencing unrelenting drought since 2004. An estimated 41 percent of Australia's agriculture is being hit by the worst drought in the 117 years that records have been kept. The devastating firestorm of February 2009, though it had an arson component, was largely a consequence of the extreme dryness of the area.

In the drought-affected regions of the Middle East and Central

Asia, total wheat production declined by more than 22 percent in 2009. Major reservoirs in Turkey, Iran, Iraq, and Syria are at low levels, and irrigation supplies from reservoirs, rivers, and groundwater have been critically reduced. Europe, the only major agricultural region unaffected by prolonged drought, may face a drop in food production due a combination of other factors, such as late plantings, impoverished soils, reduced inputs, and light rainfall.

Low food reserves make the world's falling agriculture output particularly worrisome. The combined averaged of the stock levels of the major food exporting countries—Australia, Canada, United States, and the European Union—have been steadily declining. In the period 2002–2005 the combined stocks amounted to 47.4 million tons, in 2007 they dropped to 37.6 million tons and in 2008 to 27.4 million tons. At these levels the reserves are not sufficient to cover the import needs of countries suffering from drought and other forms of climate change.

Biologist James Lovelock's assessment of the planet's ecological condition has an ominous ring of truth. "I now take an apocalyptic view of the future" he said, "because I see 6 to 8 billions of humans faced with ever diminishing supplies of food and water in an increasingly intolerable climate."*

UNSUSTAINABILITY IN THE ECONOMY

Economy, in its original meaning, is "the management of resources for the household" (from the Greek *oikonomia,* where *oikos* is "household," and *nemein* is "manage"). The global economy is thus the system concerned with the management of the resources of the household of humanity. Not surprisingly, it faces a crisis.

The overexploitation of resources. The rising curve of human demand is beginning to exceed the descending curve of global supply. This is

---

*James Lovelock, *The Revenge of Gaia* (London: Allen Lane, 2006).

unprecedented. For most of history, humanity's demand has been insignificant in relation to the available resources. Even if local resources have been occasionally overexploited—today's arid Middle East was known as the Fertile Crescent in biblical times, and the Iberian Peninsula was a verdant region in the early Middle Ages—people could conquer new territories and find fresh resources. Today, there are no virgin territories left to conquer. In the six decades since World War II, we have consumed more of the planet's resources than in all of history prior to that time.

Human consumption is nearing planetary limits. The production of oil, fish, lumber, and other major resources has already peaked; forty percent of the world's coral reefs are gone, and annually about 23 million acres of forest are lost. Ecologists now speak also of "peak water," suggesting that the availability of water suited for human use will shrink, and its cost will increase.

Not the sheer size of the human population is the problem, but its per capita resource use. According to the UN Environment Programme's fourth Global Environment Outlook (GEO-40, October 2007), average resource demand is currently around 8.9 acres (3.92 hectares) per person. (Actual demand varies greatly: it ranges from 1.23 acres in the poorest countries such as Bangladesh, to 25.5 acres in the United States and the oil-rich Arab states.) However, the amount of land that could sustainably satisfy human requirements, the "Earth-share" of every woman, man, and child, is 4.2 acres (1.85 hectares).

The problem of reducing the 8.9 acres global average to the sustainable Earth-share of 4.2 acres is compounded by the rapid growth of the population. World population has increased from about five billion twenty-two years ago to nearly seven billion today. Since the amount of available land remains constant—and is even shrinking due to overpaving and erosion—the per capita availability of land for meeting human requirements has already shrunk from 19.5 acres per person in 1900 to 5 acres in 2005. Today only the optimum use of every square meter of available land could satisfy the basic needs of the human population on a sustainable basis.

Unsustainability in the financial system. The instability of the world's financial system is a major element in the cluster of economic unsustainabilities. This instability has been growing over the years, but until the 2008 financial crisis it was not known beyond economists and analysts. Those who knew it firsthand were those who profited from it. They didn't take determined steps to change the system: they were hoping that it would continue to produce wealth for them.

The already unstable world economic system was badly shaken by the burst of the U.S. sub-prime bubble. This had its roots in the low interest rates the Federal Reserve created to accommodate the aftereffects of the burst of the previous Internet bubble; it led to an unprecedented and largely unanticipated boom in the U.S. housing market. Ever more houses were sold, at ever-higher prices, and with ever more profit for banks, brokers, and the whole financial sector.

When the number of people who could afford to buy houses at the inflated prices dropped, ingenious mortgage constructions were put together to enable people who couldn't really afford to buy real estate to stay in the market. Mortgages were offered at attractive low interest rates, but the lenders could adjust them upward. People were allowed, and in some cases even encouraged, to lie about their finances when filling out the applications. The lenders knew that many people would default on their payments, but they continued to entice prospective buyers to sign the mortgages for the profits they obtained from them.

This shortsighted practice was bound to crash. When it did, it led to the almost instant loss of over two million jobs in the United States alone. Worldwide it produced the greatest loss of wealth ever recorded apart from a major war: 2.8 trillion dollars. When the losses on the world's stock markets are added, the total cost of the crisis, according to the Bank of England, rises to an astronomical 26 trillion dollars.

The structural unsustainability of the world's financial system goes deeper than the creation and bursting of speculative bubbles. This unsustainability is likewise due to an irresponsible search for short-term benefits without regard for the long-term consequences, and it involves the balance of trade of the major players. For the past several decades the United States has been running up a massive international debt to finance unrestrained domestic consumption. The excess of imports over exports has produced a staggering trade deficit with China and other newly industrialized Asian economies. The latter have exported more than they imported, and chalked up a large trade surplus. Currently the central banks of the Asian economies are financing U.S. overspending, but they are looking for ways to free themselves from captivity by America's fiscal policy.

The IMF's *Economic Outlook* noted as early as 2005 that it's no longer a question of *whether* the world's economies will adjust, only *how* they will adjust. If merely more money is pumped in without basic structural reform, the adjustment will be "abrupt," with hazardous consequences for global trade, economic development, and international security.

With the 2008 financial crisis abrupt change has already happened, and it has affected the entire world's economic-financial system. How that system will change to avoid collapse is too early to tell, but it's already clear that it will never be the same again.

UNSUSTAINABILITY IN SOCIETY

We now look at the cluster of unsustainabilities in society.

The rich-poor gap. There are nearly 800 billionaires in the world; during 2008, 178 were added in the United States alone, although fifty years ago there was not a single one. Eighty percent of the world's domestic product now belongs to one billion people, and the remaining twenty percent is shared by almost six billion.

Poverty has not diminished in absolute numbers. Of the nearly seven billion who now inhabit Earth, the World Bank estimates that 1.4 billion live on less than 1.25 dollars a day and an additional 1.6 billion on less than 2.50 dollars. In the poorest countries seventy-eight percent of the urban population subsists under life-threatening circumstances: one in three urban dwellers live in slums, shantytowns, and urban ghettoes, and more than 900 million are officially classified as slum-dwellers.

The population of the poor countries is increasing. If current trends continue, by the middle of the twenty-first century more than ninety percent of the world's peoples will inhabit them. The population of the poorest countries will increase from 800 million today to 1.7 billion in 2050. Population would triple in Afghanistan, Burkina Faso, Burundi, Chad, Congo, East Timor, Guinea-Bissau, Liberia, Mali, Niger, and Uganda. At the same time, the population of the industrialized countries will either shrink or remain constant.

The growing gap between the rich and the poor is a major source of unsustainability in the world. The gap is not just an economic statistic; it's a critical social reality. It depresses the life, and even the chances of survival, of the poor, while it places wealth in the hands of the rich beyond any possible use in satisfying the real requirements of their life.

Poverty is not due to the inadequacy of the planet's physical and biological resources: it's due to the faulty functioning of the system that processes and distributes them. With a better use and distribution of the available resources all people could achieve a decent standard of living. Agriculture and trade could evolve more equitably, and every person could obtain the 1,800 to 3,000 calories required for health. As it is, people in North America, Western Europe, and Japan use (and to some extent waste) 140 percent of their daily caloric requirement, whereas populations in countries such as Madagascar, Guyana, and Laos subsist on 70 percent.

Current trends in energy consumption reflect the polarization of society along the rich-poor divide. The average amount of commercial electrical energy consumed by Africans is half a kilowatt-hour (kWh) per person; the corresponding average for Asians and Latin Americans is 2 to 3 kWh, and for Americans, Europeans, Australians, and Japanese it's 8 kWh. The average American burns 5 tons of fossil fuel per year, in contrast with the 2.9 tons of the average German.

Energy use correlates with air pollution. The average U.S. household produces roughly 150 pounds of CO_2 a day, which is twice as much as that produced by the average European household and five times the global average. Altogether the average American places twice the environmental load of the Swede on the planet, three times that of the Italian, thirteen times the Brazilian, thirty-five times the Indian, and two hundred and eighty times the Haitian. These conditions polarize the population of the planet. They feed resentment and animosity and breed violence. They are humanly intolerable and socially unsustainable.

The breakdown of social structures. Social structures are breaking down all over the world. Frustrated and marginalized sections of the population are becoming desperate and aggressive. Mounting stress and intolerance pits ethnic and faith communities against each other. Except for the United States, where Barack Obama's election generated a fresh wave of hope and confidence, the political leadership is cut off from the mainstream of society and is mistrusted. The same applies to the world of big business and big finance. As malaise and discontent grow, those who oppose the dominant system are turning increasingly radical.

In poor countries the struggle for economic survival destroys the traditional extended family. Women are extensively exploited, given menial jobs for low pay. Fewer women than ever have remunerated jobs and more are forced to make ends meet in the "informal sector." As they are obliged to leave the home in search of work, poverty breaks apart even the nuclear family.

Children fare even worse. According to the International Labour Organization, fifty million children are employed for a pittance in factories, mines, and on the land, for the most part in Africa, Asia, and Latin America. Many more are forced to venture into the hazards of life on the street as "self-employed vendors"—the euphemism for child beggars.

Social structures are breaking down in the rich countries as well. Job security has become a thing of the past; competition is intensifying, and the gap is growing between the rich and the poor. As people are stressed, family life is suffering in the affluent populations as well.

In the U.S. the rate for first marriages ending in divorce is fifty percent, and about forty percent of children grow up in single-parent families for at least part of their childhood. Less than 50 percent of U.S. couples now live with a child in the home. More and more find satisfaction and companionship outside rather than within the family. This is the case especially after children have "flown the nest." It's becoming usual for couples to seek fulfillment with other partners rather than restructuring the family relationship in a childless situation.

Many of the functions of family life are taken over by outside interest groups. Child rearing is increasingly entrusted to kindergartens and company or community day-care centers. The provision of daily nourishment is shifting from the family kitchen to supermarkets, prepared food industries, and fast food chains. Leisure-time activities are strongly colored by the marketing and public relations campaigns of commercial enterprises.

Families eat meals together less and less frequently, and when they do, the TV is likely to be the center of attention. Children's media exposure to TV, video games, and "adult" themes is increasing, and exposure to such imagery, researchers find, connects with violent and sexually exploitive behavior. Teens face the peer challenge of "freer" sex, where loose "hooking up" for one-night stands is seen as normal, and building deep emotional relationships with sexual partners is out of date.

A Chinese proverb warns, "If we don't change direction, we will end up exactly where we are headed." In our ecologically, economically, and socially unsustainable world, this would be disastrous.

THREE

The 2012 Horizon

THE *TIME* IN "TIMELY CHANGE"

We need a major shift: this world is not sustainable. But how much time do we have for a shift?

Not much. The window of time for meaningful change is closing faster than anybody had thought. There are two principal reasons for this. One is the unexpected acceleration of current trends, and the other the widely neglected but critical feedbacks and cross-impacts that exist among the trends.

UNEXPECTED ACCELERATION

Current trends are building unexpectedly fast toward points of irreversibility. As a result, the time estimates of when these critical "points of no return" will come about have shrunk from the end of the century to mid-century, then to the next twenty years, and—for some trends—to the next five to twenty years.

- The global sea level has been rising one and a half times faster than predicted in the Third Assessment Report of the IPCC (Intergovernmental Panel on Climate Change), published in 2001. Forecasts published at the end of 2008 project a global sea level rise that is more than double the 0.59-meter rise forecast even by the Fourth Assessment Report. The rise is fed by the rapid melting of glaciers in the Antarctic. In February of 2009 the Scientific Committee on Antarctic Research reported that two major glaciers are sliding into the sea. Since this process involves ice that was on land rather than floating in the sea, it will create a significant rise in world sea levels. As Colin Summerhays, director of the Committee, remarked, these findings are "unusual and unexpected."

According to data presented in March of 2009 to a meeting of 2,500 researchers and economists in Copenhagen, the sea could rise by over a meter all over the world, with huge impact on the life of hundreds of millions. Lord Stern, the author of a landmark 2006 report on the economics of climate change, warned that there would be hundreds of millions, probably billions, of people who would have to move, creating decades or centuries of conflict around the world.

- Carbon dioxide emissions and global warming have likewise outpaced expectations. The rate of increase of CO_2 emissions rose from 1.1 percent between 1990 and 1999 to over 3 percent between 2000 and 2004. Since 2000 the growth-rate of emissions has been greater than in any of the scenarios used by the IPCC in both their Third and Fourth Assessment Reports.
- The warming of the atmosphere has progressed faster than expected as well. In the 1990s forecasts spoke of an overall warming of maximum three degrees Celsius by the end of the twenty-first century. Then the time-horizon for this level of increase was reduced to the middle of the century; presently some experts predict that it could occur within a decade. At the same time, the prediction for the maximum level of global warming rose from three to six degrees. The difference is not negligible. A three-degree warming would cause serious disruption in human life and economic activity, while a six-degree warming would make most of the planet unsuitable for food production and large-scale human habitation.

CROSS-IMPACTS AMONG THE TRENDS

Unexpected acceleration is one reason for the reduced time-horizon; the disregard of feedbacks and cross-impacts is another. Most predictions take only one trend into consideration, such as: global warming and attendant climate change; water quality and availability; food production and self-reliance; urban viability, poverty, and population pressure; air quality and minimal health standards, or others. They fail to consider the possibility that a critical point reached in one trend could drive other trends toward a point of no return. Yet there are multiple feedbacks and cross-impacts among the global trends, both in nature and in the human world. For example:

- An increase of greenhouse gases in the atmosphere leads to global warming, and warming creates drought, thereby reducing the planet's vegetation cover. That, in turn, limits the biosphere's capacity to absorb carbon.
- Warmer water in the oceans melts polar ice and threatens to alter the course of major ocean currents, such as the Gulf and the Humboldt. The altered course of the streams provokes further changes in the climate.
- A drop below the minimum required for health in the quality of the air in urban and industrial megacomplexes leads to a breakdown in public health, with serious social, economic and political implications.
- The continued warming of the atmosphere produces more and more drought in some areas and coastal flooding in others. Starving and homeless people are impelled to move from the impacted areas to less hard-hit regions and create critical food and water shortages and social conflict in vast regions.

The impact of one trend on another reduces the time available for effective change. There is now a distinct probability that one or another

The 2012 Horizon   25

vital trend will reach a critical point within four or five years. When that happens, the chain-reaction catalyzed by it will engulf not only the immediately affected region and population, but entire continents.

The bottom line is that the time left for averting a global breakdown is perilously close to the famed 2012 prophesies of the end—or if not the end, then the transformation—of the world.

THE 2012 PROPHECIES

What is it about the year 2012 that makes it a plausible point of no return for humanity as a whole? There are a wide variety of forecasts, predictions, and prophecies that speak to this possibility, some esoteric, others scientific. The most widely cited are the prophecies that flow from the Mayan calendar.*

The Mayan calendar was completed by priest-astronomers in the year 1479, and was carved into the famous Aztec-Mayan sun stone. It details immense passages of time, and includes mathematical calculations so accurate that modern astronomers are at a loss to understand how they were carried out. The most immediately relevant time-calculation is the so-called long count. This marks the end of the Fourth Sun—also known as the Fourth World—on the 21st of December 2012 (the date is actually the 28th of October 2011, but counting the accumulated time-shifts yields the 2012 date).

The long count doesn't say that with the end of the Fourth World the world itself will come to an end. Carlos Barrios, a Guatemalan historian and anthropologist who became a Mayan Ajq'ij (ceremonial priest and spiritual guide), is a spokesperson for the Mayan elders. He is definite on the question regarding the end of the world. "Other people write about prophecy in the name of the Maya," he declared in a series of interviews in Santa Fe. "They say that the world will end on December

---

*For the fundamental implication of these prophecies, see the afterword, by Jose Argüelles.

2012. The Mayan elders are angry with this. The world will not end. It will be transformed." Everything will change, Barrios said. December 21, 2012 will be a date of rebirth, the beginning of the World of the Fifth Sun. If the people of the earth can reach this date without having destroyed too much of the earth, they will rise to a new, higher level.*

The Mayan is not the only culture that foresees a radical transformation at the end of 2012. For the first time in 26,000 years, at sunrise on December 21, 2012 the Sun will rise to conjunct the intersection of the Milky Way and the plane of the ecliptic. This cosmic cross is seen as a fundamental re-alignment by a number of spiritual traditions, including the Hopi time-keeping system, Vedic and Islamic astrology, Mithraism, the Jewish kabala, the cycle of yugas in Hinduism, European sacred geography, medieval Christian architecture, and a variety of hermetic metaphysics.

Astrological calculations yield the same date. They show that on December 21, 2012 the center of our galaxy will complete a "cosmic year": a 25,920-year journey around the wheel of the zodiac. According to astrologers a new cosmic year will then begin, lasting for another 25,920 of our years.

A mathematically elaborated "timewave" developed by mystical philosopher Terence McKenna, on the basis of an interpretation of the I Ching, also points to the 2012 date. The timewave computes the ebb and flow of novelty in the universe, considered an inherent feature of time. The graph that maps the wave shows when, but not where, "novelty"—newness, or more exactly, the density of evolutionary complexification and dynamic change—increases or decreases in the history of Earth.

The great periods of novelty began about four billion years ago when the planet was formed, and continued sixty-five million years before our time when the dinosaurs became extinct and mammals dif-

---

*For this citation and more information on Barrios and the prophecies of the Maya see Saq' Be': Organization for Mayan and Indigenous Spiritual Studies, www.SacredRoad.org.

fused over the continents. There was a surge of novelty between 15,000 and 8,000 B.C.E., the approximate period of the Neolithic Revolution and the birth of agriculture with settled communities. Another surge occurred around 500 B.C.E. when Lao-Tzu, Plato, Zoroaster, Buddha, and other seminal figures appeared on the scene.

According to the graph each period of novelty occurs sixty-four times faster than the one before. The calculation yields modern-age novelty waves in the late eighteenth century, coinciding with the epoch of social and scientific revolutions; during the unsettled 1960s; and in the early twenty-first century, around the period of the 9/11 terrorist attack. The next peak of novelty fell in November of 2008, the date of the transformative U.S. presidential elections, and another wave is to occur in October of 2010.

The timewave will culminate on 21 December 2012. At that point novelty on the planet will reach infinity. This would be an endpoint—a time at which anything and everything conceivable to the mind would occur at the same time.

A prediction coming from astronomy, a "hard" physical science, coincides with the date given by the esoteric traditions and the mathematical calculations based on them. Astronomers have noted that since the 1940s, and particularly since the year 2003, the Sun has become remarkably turbulent. Solar activity is predicted to peak around 2012, creating storms of an intensity unprecedented since the 1859 "Carrington event," when a large solar flare accompanied by a coronal mass-ejection flung billions of tons of solar plasma into Earth's magnetosphere. It disrupted Victorian-era magnetometers and the world's telegraph system. The next solar maximum, expected to occur in 2012, would do much greater harm, as human life and civilization have become strongly dependent on the globally interlinked energy grid. According to "Severe Space Weather Events: Understanding Economic and Societal Impacts," a National Research Council report issued in the spring of 2009 by the U.S. National Academy of Sciences, another Carrington event would induce ground currents in the U.S. that would knock out three hundred key transformers within 90 seconds and would cut off power for more than 130 million people. Its cost could be as high as 2 trillion dollars, and recovery would take between four and ten years. An even worse impact would be felt in China, where the electrical grid is more vulnerable than in the U.S. and other industrialized countries.

Another forecast coming from the physical sciences concerns the behavior of Earth's magnetic field. This field appears to diminish in intensity, and produce vast holes and gaps. At the end of 2008 NASA's *Themis* spacecraft discovered a hole ten times larger than previously estimated: it is four times the size of Earth. A pole-shift, the reversal of the North and South magnetic poles, is possible. It, too, could occur in a matter of years, possibly by the end of 2012.

We have no certainties but, as we have seen, a series of curious "coincidences." It's not certain, but more than likely, that the period around the end of 2012 will be turbulent. In the esoteric perspective this period will mark a major phase-change in human affairs, with the pessimistic views foreseeing the end of the world and the optimistic interpretations anticipating the advent of a new world. In the scientific perspective we can expect another CME (coronal mass ejection) by the Sun, altering the physics of Earth and impacting human life and economic activity throughout the planet.

None of the prophecies and predictions is one hundred percent certain, but in their ensemble they are highly significant. When we also take the time-horizon given by the cross-impact of global trends into account, we get serious grounds for viewing the end of 2012 as a critical point in history, when the fate of humankind could hang in the balance.

Part Two

WAKING UP—IN TIME

We are now faced with the fact . . . that tomorrow is today. We are confronted with the fierce urgency of now. . . . Over the bleached bones and jumbled residues of numerous civilizations are written the pathetic words—"too late."

MARTIN LUTHER KING
SPEECH AT RIVERSIDE CHURCH
NEW YORK CITY, APRIL 4, 1967

FOUR

The Timely Change Objectives

We face a choice of destiny. We must be clear that we face it, and also that we can do something about it. We have a number of alternatives available to us.

- We can try to carry on as before and hope for the best.
- We can try to make whatever changes seem personally, economically, and politically expedient.
- Or we can move decisively toward conscious and timely change.

The rational and responsible choice is to prepare the ground for timely change—before it's too late. This calls for conscious, purposive change, by each government, each business, and each person. It begins with the reassessment of the objectives that now orient our life and will soon decide our future: the objectives of politics and business—and the personal objectives to which we ourselves subscribe in life.

THE OBJECTIVES OF A NEW POLITICS

The objectives of an enlightened politics embody the perennial principles of a democratic society: to further the interests of the *demos*, the people. Acting on these principles is timely and realistic. A new leadership has been taking the helm in the United States and in a number of other countries, and it's more open to serving the true interests of the people than the old guard.

The true interests of the people include physical survival, stable relations in society, a meaningful social and cultural identity, and remunerated and socially useful work. Safeguarding these interests calls for conditions where the necessary resources are physically available and economically accessible to everyone.

> *Physical survival security* calls for access by all members of the community to the basic resources of life: adequate supplies of food, water, shelter, and clothing.
>
> *Ensuring stable interpersonal relations* calls for social and economic conditions suitable for maintaining family life, and functional and beneficial relations within the community.
>
> A *meaningful social and cultural identity* requires a judicial system dedicated to social and economic justice, and a system of education and information that helps people recognize themselves as unique yet integral parts of a possibly multiethnic and multicultural community.
>
> *Remunerated and socially useful employment* requires in turn that the economy is maintained at a level where it can provide jobs for all people willing and able to work.

In a democratic society the political leadership consults the people in translating these objectives into practical policies. In the interest of a fair and inclusive consultative process it fosters gender equality, and ensures that all sectors of society can express their needs and requirements and participate in the decisions that respond to them.

Beyond these basic objectives, the goals of a new politics are specific to our time. In today's world, enlightened political objectives focus on the economic, social, and ecological sustainability of society. The political leadership does the following:

- estimates, takes into consideration, and publishes the carbon footprint of activities in the public sector
- uses public funds exclusively for organic, sustainable, and fair-trade products and services
- requires businesses to measure the carbon footprint, recycle-ability, toxicity, and bio-degradability of their products and to display the corresponding index on their products and in their marketing
- provides tax and other economic incentives for the use of alternative energy and resource-saving and recycling technologies, and technologies of low- or zero-waste emission
- works with local businesses to increase the durability and the energy and resource efficiency of their products
- pays attention to the accessibility for all people to natural common goods, including energy, water, and land
- improves the quality and increases capacity of the public transportation system, creating realistic alternatives to the use (and overuse) of the personal automobile
- counteracts tendencies toward excessive consumerism by limiting the advertising of nonessential (and not only of toxic) goods in public spaces and in the media
- channels funds to reconstruct and revitalize derelict or disadvantaged areas
- uses safe and efficient alternative energy technologies in electric power generation, transport, and communication
- monitors, and as necessary regulates, civil and industrial activities that threaten to destroy ecological balances and despoil or reduce wilderness areas
- applies strict criteria for urban design and construction, requiring renewable energy technologies and efficient insulation to be part of public housing and a precondition of licensing the construction of private dwellings, commercial buildings, and industrial plants

- makes accessible fields, forests, rivers, streams, lakes, and seas in the surroundings, with adequate provision for the integrity of ecological cycles and processes

GRASSROOTS SUPPORT FOR THE OBJECTIVES

There are trends and movements in society that are in line with such policies. A new culture is emerging at the creative edge and is moving rapidly toward the mainstream. It's the culture of the "cultural creatives"—a term introduced by public opinion researcher Paul Ray. Their value shift embraces a shift in lifestyles, from matter- and energy-wasteful ostentation toward ways of living marked by voluntary simplicity and a search for harmony with nature.

Cultural creatives are dedicated consumers of intense, enlightening, or enlivening experience such as weekend workshops, spiritual gatherings, personal growth events, and experiential vacations. They view themselves as synthesizers and healers, on the personal, the community, and even on the planetary level. They are committed to holistic ways of thinking and acting, eating natural whole foods, turning to holistic health care, searching for holistic inner experience, and seeking a holistic balance between work and play, consumption and personal growth.

Ray's survey of March of 2008 found that most Americans are ready to embrace action for the greater good of society, even of the planet. There is a growing awareness of the need for action. Most people recognize that we are in the midst of a climate-change crisis, and they see it as a sign of a full-blown planetary emergency. Some 87 percent of the more than one thousand individuals who responded to the 500 questions in Ray's survey agreed that "we need to treat the planet as a living system"; 81 percent agreed that "corporations must take more responsibility for their impact on global warming"; 75 percent agreed that "people need to work for the good of the planet, for it is our only home"; 70 percent agreed that "I see myself as a citizen of Planet Earth as well as an

American"; and 68 percent were in agreement that "at this time in history we need to see this as all one planet and one humanity."*

There is awareness of the need for change especially in regard to the environment. More than 62 percent agreed that "the earth is headed for an environmental catastrophe unless we change"; 56 percent agreed that "our materialistic way of life can be replaced by a new more hopeful one"; and 51 agreed that "I am willing to do volunteer work as part of a commitment to help save the planet."

In a related finding, the Institute of Noetic Sciences found that in a significant segment of the U.S. population, thinking is changing: from competition to reconciliation and partnership; from greed and scarcity to sufficiency and caring; from outer to inner authority (from reliance on outer sources of "authority" to inner sources of "knowing"); from mechanistic to living systems (from concepts of the world modeled on mechanistic systems to perspectives and approaches rooted in the principles that inform the realms of life); and from separation to wholeness, brought about by a fresh recognition of the oneness and interconnectedness of all aspects of life and reality.

Similar changes are occurring in other parts of the world. A survey carried out in 2005 by the Italian branch of the Club of Budapest found that 35 percent of adult Italians live and act as cultural creatives. Analogous figures are coming to light in surveys in other countries of Europe, as well as in Japan, Australia, and Brazil.

Entrepreneur and environmentalist Paul Hawken estimated that there are now one million social change organizations in the world, ranging from small neighborhood associations to well-funded and relatively powerful NGOs. They share a common set of values and collectively comprise the largest social movement in history. This, Hawken noted in his 2007 book *Blessed Unrest: How the Largest Movement in the World Came into Being and Why No One Saw It Coming*, is

---

*The poll was conducted by the Institute for the Emerging Wisdom Culture at Wisdom University. See www.wisdomuniversity.org/cultural-creatives.htm.

not the standard "social movement," for ordinarily movements have leaders and doctrines. The new movement is dispersed, inchoate, and fiercely independent. It testifies to the emergence of what David Korten, Nicanor Perlas, and Vandana Shiva called "global civil society"—the "social expression of the awakening of an authentic planetary culture grounded in the spiritual values and social experience of hundreds of millions of people."*

Cultural change in society is more rapid and powerful than politicians as yet realize. A critical mass is not only ready and able to elect leaders who stand for change, it's also ready and able to provide active support for enlightened policies.

THE SOCIAL OBJECTIVE IN BUSINESS

Business is said to be the private sector, but it has become so powerful that it's no longer a "private" sector—it's the most public sector there is. The top five hundred industrial corporations employ only 0.05 percent of the human population but control 70 percent of world trade, 80 percent of direct foreign investment, and 25 percent of world economic output. Even if reduced by the economic crisis, the sales of the largest companies, such as Toyota, Ford, Mitsui, Mitsubishi, Royal Dutch Shell, Exxon, and Wal-Mart, still exceed the GDP of dozens of countries, including Poland, Norway, Greece, Thailand, and Israel.

Wielding unparalleled wealth and power, business companies are a key factor in the equations that decide the human future. States and governments, the "public sector," can at best regulate their activities, and their power to do even that is limited. When the public sector sets up too many regulations, private-sector multinationals move elsewhere.

During the past fifty years or more, the private sector has progressively detached itself from the main body of society, with companies

---

*David Korten, Nicantor Perlas, and Vandana Shiva, *Global Civil Society: The Path Ahead* (People-Centered Development Forum, www.pcdf.org/civilsociety/default.htm).

pursuing their own ends of profit and growth. If this sector is to be reintegrated in society, there must be a fundamental shift in the objectives to which companies are dedicated.

The recognition that the shift required of business is fundamental and not just tactical has not fully penetrated the thinking of managers. Until recently a cautious optimism was spreading in the corporate world, based on the assumption that companies could pull out of the morass of economic stagnation and ecological degradation without having to change in a fundamental way. In that context, concern with the environment is only an additional frill that can be jettisoned when the "serious" problems of economics and finance demand attention. Business can remain basically as usual, only instead of focusing on "new and improved" products and services—on things that are more powerful, more sophisticated, and more attractively designed and packaged—companies need to focus on products and services that are "sustainable" and "environmentally friendly." This goes along with the attitude that if the client or consumer wants more sustainability, such features should be introduced into the service or product palette. If he wants more organic and environment-friendly products, do likewise. This will work as long as the client or customer doesn't need to make sacrifices. Paying more, or making do with a product that's less powerful, sophisticated, or attractive-looking, are perceived as sacrifices: they are to be avoided. The winning tactic is to produce products and offer services that do the same as they always have, but have additional features that permit their being marketed as sustainable and eco-friendly.

The global financial crisis has shed serious doubt on the assumption that this tactic can work. Business as usual with superficial modifications is unlikely to cope with the degradation of the environment and the instability of the world's economic and financial system. Danish Prime Minister Anders Fogh Rasmussen has said that business as usual is dead—green growth is the answer to both our climate and economic problems. And economist David Korten has pointed out that applying the kind of thinking that has led to the crisis to attempts to resolve it is the height of insanity.

The classical business objective. The classical objective to which companies dedicate themselves was stated by economist Milton Friedman in a frequently cited 1970 article in *The New York Times Magazine*. The only legitimate objective and rational purpose of a business company, wrote Friedman, is to make money for its owners. Everything else is secondary and incidental.

This is the "shareholder philosophy." There was little recognition of a viable alternative to it in the 1970s, and there was not much awareness in the business world of the need for one. This is no longer the case. Today business leaders know, even if many do not act on it, that the responsibility of the company extends beyond its shareholders, to the people whose lives are touched by its activities. These are the *stakeholders.* A person is a stakeholder whether he or she is an employee, a customer, a client, or simply a member of the host community. The lives of all these people are affected by the way the company extracts, transports, and processes its resources, and discards its wastes.

While the "stakeholder philosophy" is now widely known, in practice it remains subordinated to the shareholder philosophy. In the prevalent management philosophy, assuming responsibility for the stakeholders is seen mainly as a means to an end: the end remains maximizing profit for the shareholders. Assuming a higher level of responsibility for the stakeholders seems like a good way to achieve this end: it ensures customer and client loyalty, a higher level of public approbation, and can lead to improved competitiveness and profitability.

Shifting beyond the classical objective. While the main body of the business world is slow to change, there are business leaders and entire companies that are willing to shift beyond the classical objective. This was shown by an event reported by David Cooperrider, founder and chairman of the Center of Business as an Agent of World Benefit. A $10-billion-year multinational, near the top of Barron's list of 500 Best

Performing Companies, called together its executives to focus on the "ten largest global problems facing humankind." This strategy summit, kept secret for proprietary and competitive advantage reasons, confronted the executives with the question: how can the ten largest problems of the world be turned into strategic business opportunities for the company— innovations in products and operations, opening new markets, igniting customer loyalty, reducing risk, bringing down costs—and how could the company at the same time contribute to building a more secure and better world? By the end of the three-day meeting a new vision of the company's future began to take shape: as a solution provider to the ten largest global challenges facing humankind. The summit, though called to serve the classical business objective, led to a deep-seated transformation in the goals and mission of the company.*

Another example of a deep-seated shift in the objective of business is the Grameen Bank, founded by Mohammad Yunus. This bank, discussed in his 2007 book *Creating a World Without Poverty: Social Business and the Future of Capitalism,* specializes in making small loans—"microcredits"— to poor people, mainly women, without requiring collateral. It embodies what Yunus calls "social business"—a way to use the power of business to tackle a wide spectrum of social problems, from poverty and pollution to inadequate health care, lack of education, and hunger.

Yunus, who in 2007 was awarded the Nobel Peace Prize for his initiative, foresees a world transformed by thousands of social businesses. Following his meeting with Danone's chief executive Franck Riboud in 2005, a first such social business was created: Grameen-Danone Foods of Bangladesh. Its objective is to supply nutritious food to poor children in one of the poorest countries in the world. More recently another social business was born in Bangladesh: Grameen Shakti. Up until now it has installed 160,000 solar units in poor households, and by 2012 the number is expected to grow to 7 million.

---

*See the postscript to Chris Laszlo, *Sustainable Value: How the World's Leading Companies Are Doing Well by Doing Good* (Sheffield: Greenleaf Publishing, 2008).

The unfolding of a worldwide wave of social businesses would be a promising step toward bringing business back into the fold of socially committed actors. But if the majority of companies, including global-market leaders, remain committed only or mainly to their own short-term interests, a wave of social businesses unfolding in parallel with businesses committed to the classical objectives would not be sufficient: it would be a case of too little, too late. The worldwide economic and ecological crisis would reach a critical phase before the combined weight of socially committed companies could provide a meaningful counterweight.

The timely transformation of a critical mass of major companies remains a fundamental requirement. For the present it is blocked by the dominant business ideology. If socially committed executives do not bring in sufficient profit to remain competitive in short-term oriented markets, they are likely to be dismissed and replaced by more "realistic" managers.*

Prospects for a sustainable shift. A sustained change in the business world requires a change in the thinking of those who control major segments of the market. If the market leaders recognize that it's in their own interest to initiate a timely shift, they could sustain it. They could form alliances among themselves to ensure that socially and ecologically responsible practices become the norm in their sector of industry.

Public-spirited alliances could find support among business leaders. Although the ownership of most companies is anonymous and only concerned with making money, some companies remain controlled by individuals and families who conserve a sense of mission about their companies' activities. They are the contemporary equivalents of the legendary "captains of industry" of the first decades of the twentieth century. A Rockefeller, a Vanderbilt, a Ford, a Mellon, an Astor, and

---

*A firsthand example of this is recounted by former Sanyo chief executive Tomoyo Nonaka in the "Sanyo Experience," on page 42.

a Carnegie didn't think of himself purely as a businessman, out to get the maximum money for himself and his family; he considered himself a builder of society, a force for the common good. As IBM founder Thomas J. Watson Sr. said, companies were not created "just to make money" but to "knit together the whole fabric of civilization."

This spirit is not extinct today, and its flame could be rekindled. A Bill Gates, a Warren Buffett, and other business leaders still create and administer charitable Foundations to champion their preferred causes, much like Rockefeller, Ford, Carnegie, and others did before them. But times have changed, and this alone is not sufficient. In the 1920s and 1930s nobody suspected that a company's pursuit of business as its business would have negative consequences in society. Obviously, society needs motorcars, gasoline, steel, and the other products and services provided by the companies. For business people, being public spirited didn't involve changing the orientation of their company; it meant ensuring fair treatment for workers and staff, and espousing selected social causes on the side.

But today it's not enough to "do good" as additional philanthropy while focusing single-mindedly on "doing well" in the marketplace. The damage done by companies that persist in the pursuit of short-term profit-maximizing strategies is not made good by funding charitable causes, however important they may be. The need is for those who have the wealth and the power to control major companies to become forces for the public good not by philanthropy, but by reorienting their companies.

The social objective serves the shared interest of business and society. It's not an arbitrary leap but a consistent development of the stakeholder philosophy, rooted in the recognition that in the final count the company's stakeholder is society as a whole. Embracing it would be a truly "giant step for mankind," for it would bring the private sector into the fold of societal actors committed to the public good.

Grassroots support for the social objective. The business world is strongly dependent on the market, and the market is highly sensitive to demand and thus to approbation by the public. If people find it acceptable to settle for the same behavior by business as always, with a measure of sustainability and ecological features thrown in, then companies will do well in the marketplace with the tried and tested methods and will not see the need for change. But if peoples' thinking changes, so will the pattern of demand in the market. Intelligent executives and innovative companies can change with it.

People's thinking is in fact changing. Socially and ecologically conscious values are surfacing in wide strata of the population, and they have a marked impact on people's consumer and market preferences. In the United States the market for "values-driven commerce" reached $230 billion already by the year 2000; *The New York Times* called it "the biggest market you have never heard of."

Megatrends 2010 author Patricia Aburdene noted that values-sensitive conscious consumers—people who are part of the LOHAS (Lifestyles of Health and Sustainability) culture—make up a rapidly growing segment in five sectors of the economy:

> **in the sustainability sector,** including ecologically sound construction, renewable energy technologies, and socially responsible investments
>
> **in the healthy living sector,** appearing in the market as demand for natural and organic foods, nutritional supplements, and personal care
>
> **in the alternative healthcare sector,** comprised of wellness centers and complementary and alternative medical services and health care
>
> **in the personal development sector,** made up of seminars, courses, and shared experiences in the body-mind-spirit area
>
> **in the ecological lifestyle sector,** appearing in the form of demand for ecologically produced, recycled, or recyclable products, as well as ecotourism.

THE SANYO EXPERIENCE

A Contribution by Tomoyo Nonaka, Chairperson of Sanyo Electric 2005–2007, and current Chair of the not-for-profit Gaia Initiative (Japan)

As Chairperson of Sanyo Electric, the first thing I did was to refocus corporate activities with a new vision. When I took over in July of 2005, Sanyo with about 100,000 employees was one of the largest home electronics companies in the world, with more than thirty business divisions, sales of about 2,500 billion yen (25 billion U.S. dollars) and total liabilities exceeding 1,000 billion yen (around 10 billion dollars). The company's objective was to achieve an annual turnover of 10,000 billion yen (about 100 billion dollars) by 2010. Each division was to build and focus on its own branch of the business to reach its own sales target.

The business divisions operated separately. Each had its own technological know-how, a super-potential with which I had fallen in love during the previous three years when I had served as external director. I was certain that we could reach our objective to evolve ourselves if only we became aware that we were all one corporate family.

The corporate vision I created was called "Think Gaia." My objective was to create a corporation that uses its technological know-how to make products that solve the environmental problems we now face and enable people to live sustainably on this planet. I was certain that this know-how is much needed by Gaia.

In order to implement the corporate vision, I created a three-year Evolution Plan aimed at reorganizing the business portfolio and improving the corporate financial structure and conditions. The Evolution Plan defined a portfolio that included three new kinds of product lines:

1. **Blue Planet**—the product line that addresses global environmental issues
2. **Genesis III**—the product line that contributes to creating a clean-energy society
3. **Harmonious Society**—the product line that helps people live in harmony with the earth

In addition to reorganizing Research and Development in line with the new mission, I created the "Think Gaia Incubation Platform" to build cross-functional teams and activities, reporting directly to top management.

These measures broke down barriers among the various business divisions, including R&D, marketing, and manufacturing and design. People started to work together for the common goal and I began to see reflections and resonances on shining and smiling faces throughout the company. In a short time, Sanyo introduced more than ten new Think Gaia (TG) products, and the following four "world-first" products:

1. **Eneloop batteries:** rechargeable batteries that can be recharged up to a thousand times. One Eneloop battery does away with the need to dispose of one thousand dry batteries. Through collaboration among several business divisions and using their technologies, we developed a small solar station to charge the Eneloop battery, the "Eneloop Universe."

2. **Aqua:** a washing machine and dryer, developed in collaboration with two business units, that reduces water usage in a wash cycle from 200 liters to 8 liters by purifying the used water and making air-washing possible through ozone technology.

3. **Enegreen:** an innovative way to reduce electricity consumption in air conditioners, refrigerators, and food showcases in stores and supermarkets, developed in collaboration between a number of business divisions. Enegreen absorbs as much CO_2 as a forest with an area 130 times larger than the store.

4. **Virus-Washer:** likewise a collaborative development, this is an air-cleaner that eliminates up to 99 percent of airborne viruses, including avian influenza (bird flu) virus, using a technology that electrolyzes simple tap water.

While I served as Chairperson, Sanyo underwent drastic restructuring in the first two years. Under the three-year Evolution Plan the bottom line was raised from red to black, with an operating income of about 750 million dollars and a net profit of around 280 million for the fiscal year ending March 2008. However, throughout my tenure I had to fight about how to run the company with board members sent by investors. They said Think Gaia is too "naive" and "feminine" to produce the desired profit.

The conflict with investors and the board led to my resigning as Chairperson of Sanyo in March of 2007. I established The Gaia Initiative, a not-for-profit organization that sets forth my life mission. The Sanyo managers who adopted the Gaia vision continue to work for it in their new corporate environment. It is my hope and belief that the kind of social and ecological vision instilled in Sanyo under the Think Gaia plan will spread in the world, and help shift the basic objective of business.

The Italian social entrepreneur Marco Roveda built LifeGate, a highly successful organization based on the shift from "the civilization of compulsive consumption" to "the civilization of conscious consumption."

According to Roveda this means shifting

- ► from having to being
- ► from living on the shoulders of nature to being a harmonious part of the ecosystem
- ► from GM- and pesticide-based agriculture to organic agriculture
- ► from choosing work on the basis of "having" to choosing on the basis of "being"
- ► from overcoming others to giving consideration to others
- ► from living superficially to giving meaning to one's life
- ► from business is business to life is life
- ► from personal interests to the interests of the community
- ► from mutual-interest acquaintances to true friendships
- ► from exploitation to respect
- ► from conditioning to liberty
- ► from inconsiderate consumption to ethical consumption
- ► from mere formalities to sincerity
- ► from being a spectator to being a participant
- ► from money to love.*

There are similar shifts in many parts of the world. They indicate a groundswell in society of vital relevance to the conduct of business.

THE OBJECTIVES OF PERSONAL RESPONSIBILITY

In the foreword to this book, Mikhail Gorbachev wrote,

We must not wait until society's crisis reaches the danger point. We must act! Every person can act. If everyone does his or her bit, together

---

*See www.lifegate.it.

we can accomplish what is necessary. We can make an impact on those who decide the politics and the destiny of society, and motivate them to begin making the necessary changes. Changes that not only resolve the crisis, but take us on a path of survival, of healthy development for people and nature, and a better quality of life for all.

Barack Obama reaffirmed this insight.

Change will not come if we wait for some other person or some other time. We are the ones we've been waiting for. We are the change that we seek.

A change has become necessary in the way we live and think. We need to forget some old and long-cherished values and beliefs, update the ethics we live by, and do our best to evolve our consciousness—grow in the vital inner dimension of our existence.

Forget Your Old Values

The crisis affecting our economy is a crisis of our civilization. The values that we hold dear are the very same that got us to this point. The meltdown in the economy is a harsh metaphor of the meltdown of some of our value systems. . . . The only hope lies in a fundamental re-examination of the values that we have lived by in the past 30 years.

BEN OKRI IN *THE TIMES* (LONDON),
OCTOBER 30, 2008

The physicist Werner Heisenberg once remarked that the problem with physicists is not that they can't learn—it's that they can't *forget*. This problem is not unique to physicists: it afflicts everyone. There is a great need for new values, yet we still hold on to the old. This is a problem, for the values that hallmarked the last few decades are outdated. The beliefs and perceptions on which they are based are obsolete, and in some cases downright dangerous.

Some Values to Forget

1. The value of getting. In all our dealings with others, the bottom line is, what do I get out of it? The rest is icing on the cake, without real value and interest.

Mistaken. Giving is not just of as much value as taking and getting, but more. We are social beings, and doing something for others gives us pleasure whether we admit it or not. Without valuing the act of giving there would not be a real basis for community life, the same as for family life. People who just want to get from others all they can could never make a functioning community. In many traditional societies the value of giving is part of the culture. When an anthropologist asked to see the richest people in a Native American settlement, he was introduced to several people, none of whom lived very differently from the others. But when he asked about the man who lived in a luxurious house with a two-car garage, he was told: "he is not rich, he still has all his money"; the rich are those who give what they don't actually need to the community.

2. The paramount value of money. There is a direct connection between having money and being happy. The more money we have, the happier we are.

Not true. The belief about the link between wealth and happiness is not borne out by experience. Money can buy many things but not happiness and well-being. It can buy sex but not love, attention but not caring, information but not wisdom. Since 1957, the GNP in the United States has more than doubled, but the average level of happiness has declined: those who report being "very happy" are only thirty-two percent of the population. In the same period the divorce rate doubled, the teen suicide rate more than doubled, violent crime tripled, and more people than ever say they are depressed. People have big houses and broken homes, high income and low morale, secured rights and diminished civility.

3. The undiscriminating valuation of technology. Whatever the question, technology is the answer; it is useless to look beyond it. If a technology can't solve a problem today, we can develop it so it will solve it tomorrow.

Not true. Technology is a powerful and sophisticated instrument, but it's only an instrument: its utility depends on what a given technology is and how it's used. Even the best technology is a two-edged sword. Nuclear reactors produce an almost unlimited supply of energy, but their waste products and their decommissioning pose unsolved problems. Genetic engineering can create virus-resistant and protein-rich plants, improved breeds of animals, vast supplies of animal proteins, and microorganisms capable of producing proteins and hormones and improving photosynthesis, but it can also produce lethal biological weapons and pathogenic microorganisms, and destroy the diversity and balance of nature. In turn, information technologies can create a globally interacting yet locally diverse world, enabling all people to be linked wherever they live, but information networks dominated by the power groups that brought them into being flatten diversity and serve the interests only of an affluent minority.

4. *The worship of the latest and newest.* Anything that's new is better than (almost) anything that stems from last year, or the year before that.

Doesn't always hold. That the latest and the newest would always be better is not true. Sometimes it's worse than what it replaced—more expensive, less enduring, more complex, and less manageable, and more damaging to health and environment. The old is not necessarily ripe for the dust heap; it's often more enduring, more carefully produced, and more in accord with culture and tradition than the new.

5. *The fetish of efficiency.* To be successful, we must get the maximum out of every person, every machine, no matter what is produced, or whether or not it serves a humanly or socially useful purpose.

Wrong. In the contemporary world "labor efficiency" without regard to what is produced and whom it will benefit is a shortsighted strategy. It only serves the interests of corporate bosses and uncaring shareholders and leads to growing unemployment, penury, and discontent.

6. *"My country right or wrong"—the value of classical patriotism.* Come what may, I owe allegiance only to one nation, one flag, and one government.

Overstated, and misleading in practice. To a greater or lesser extent, all of us have multiple identities and allegiances. They could be extended and made more conscious. There is nothing in the healthy human mind that forbids the growth of our loyalty above the level of our state, region, or country. Americans are New Englanders, Texans, Southerners, and Pacific Northwesterners as well as Americans. Europeans are English, Germans, French, Spanish, and Italians as well as Europeans. We can be loyal to our community without giving up loyalty to our province, state, and nation. We can be loyal to our region and feel at one with an entire culture, and even with the whole human family.

The Dangerous Myths

The values we need to forget are upheld by myths that have become obsolete and actually dangerous.

The earth is inexhaustible. The idea that the earth is an infinite source of raw materials and an infinite sink of wastes is a myth, and a dangerous one at that. Its origins go back thousands of years. In the past it would hardly have occurred to anyone that the environment around them could be impoverished, its essential resources exhausted and natural cycles corrupted. When a tribe or village deforested and overworked its environment, it could move on. Today there is nowhere left to go. In a globally extended world wielding powerful technologies, the belief in the inexhaustibility of the earth leads to the unsustainable mining of natural resources and the equally unsustainable impairment of the natural cycles and balances of the biosphere.

Nature is a giant mechanism. This myth stems from the Newtonian view of the world. It was suitable for working with medieval technologies—water mills and windmills, pumps, mechanical clocks, and animal-drawn plows and carriages—but fails when it comes to living organisms and the complex world that now surrounds us. Yet most people still act as if they could engineer nature like a machine. This produces a plethora of "side effects" such as the degradation of water, air, and soil, and the alteration of the climate.

Life is a struggle where only the fittest survive. This is a myth based on a seemingly scientific concept: Darwin's theory of natural selection. It claims that in society, the same as in nature, "only the fittest survive." If you are to survive you have to be fitter for the struggle of life than others: smarter, more ambitious, more daring, richer, and more powerful. (In fairness to Darwin, we should add that by "fittest" he didn't mean the strongest and most powerful—he meant the most sensitive and capable of change.) This is another dangerous myth, for it produces widening gaps between rich and poor and concentrates wealth and power in the hands of a few clever but often unscrupulous speculators, and business and political leaders.

The market corrects economic gaps and injustices. The myth of the market is directly related to the myth of the survival of the fittest. Unlike in nature, where the consequence of "fitness" is the spread and dominance of a species and the extinction or marginalization of the others, the market myth maintains that in society there is a mechanism that distributes the benefits instead of having them accrue only to the fit. This is the free market, governed by what Adam Smith called the "invisible hand." It acts equitably: if you do well for yourself, you benefit not only yourself, your family, and your company, but also your community.

The market myth is comforting for the rich, but it disregards the fact that the market distributes benefits only under conditions of near-perfect competition, where the playing field is level and the players have a more or less equal number of chips. In the real world, the playing field is not level and the distribution of wealth is strongly skewed. Not surprisingly, the poorest 40 percent of the world population is left with 3 percent of the global wealth, while the wealth of the 793 billionaires of the world—though it plunged in 2009 to $2.4 trillion from $4.4 trillon the year before—still equals the annual income of nearly half of the world's population. It is with good reason that in his article "Common Wealth" economist Jeffrey Sachs concluded that in regard to the three great challenges of our time—ending extreme poverty, creating a sustainable environment, and stabilizing world population—relying on market forces is not sufficient.*

The more you consume the better you are. This is the popular myth that claims a strict equivalence between the size of your wallet and your personal worth as the owner of the wallet.

In the past, the equating of human worth with financial worth has been consciously fueled by business; companies did not hesitate to advertise unlimited consumption as a realistic possibility and conspicuous consumption as the ideal. Fifty years ago retailing analyst Victor Lebow said it clearly: "Our enormously productive economy demands that we make consumption our way of life, that we convert the buying and use of goods into rituals, that we seek our spiritual satisfaction, our ego satisfaction, in consumption. The economy needs things consumed, burned, worn out, replaced, and discarded at an ever-increasing rate."†

The consumption myth is not stated as brazenly as this today, but it remains powerful. In its current formulation it claims that the faster the economy grows, the wealthier we become. Economic growth, as measured by Gross Domestic Product, creates the wealth needed to provide material abundance for everyone, increasing happiness, ending poverty, and healing the environment. A rising tide lifts all boats.

The "consumption-equals-wealth-equals-happiness" myth fueled the modern consumption spree: in constant dollars humanity has consumed as many goods and services since 1950 as in all previous generations put together. Vast quantities of useful resources have been converted into toxic garbage, and the benefits have accrued only to a few rich people who reap financial gains from every transaction, regardless

---

*TIME, March 24, 2008.
†Cited in Alan Durning, *How Much Is Enough?* (New York: Norton, 1992).

of whether it serves a humanly and socially useful purpose, or depletes a resource and poisons the environment.

Economic and political ends justify military means. The ancient Romans had a saying: "If you aspire to peace, prepare for war." This made sense at the time: the Romans governed a global empire, with rebellious peoples and cultures within and barbarian tribes at the periphery. Maintaining it required a constant exercise of military power. Today the nature of power is different, but the belief about the use of war to achieve political—and also economic—objectives still persists.

But the twenty-first century is not the classical world: its social, economic, and ecological systems operate at the edge of sustainability. "Sending in the marines" can plunge a country and an entire region into social and political chaos, and because warfare devastates large tracts of land and its explosives inject vast amounts of heat into the atmosphere, war can also trigger ecological catastrophes.

Update Your Ethics

Because global interdependence demands that we must live with each other in harmony, human beings need rules and constraints. Ethics are the minimum standards that make a collective life possible. Without ethics and the self-restraint that are their result, humankind would revert to the survival of the fittest. The world is in need of an ethical base on which to stand.

UNIVERSAL DECLARATION OF HUMAN RESPONSIBILITIES,
THE INTERACTION COUNCIL, APRIL 1990

Our ethics defines our values and sets our priorities. It will be a major factor in the coming destiny choice: whether we allow ourselves to drift into a catastrophe, or head consciously toward a more peaceful and sustainable world.

Our ethics must be suited to our life and times. We live in a globally interacting and interdependent community, and our ethics cannot be a local, regional, ethnic, or national ethics—it must be a *planetary* ethics. This doesn't mean a flat, homogeneous ethics, but an ethics that draws on the humanistic elements of the world's great cultural traditions: Christianity, Islam, Hinduism, Buddhism, Shintoism, Taoism, Confucianism, and the animistic cultures of indigenous peoples.

For thousands of years, the great religions of the world offered moral precepts for their followers: codes for desirable and acceptable behavior. Examples of such codes are the Ten Commandments of Jews and Christians, and the Rules of Right Action of Buddhists. Today the power of moral injunctions based on religion has been diminished by the intellectual authority of science, yet scientists have not come up with codes for moral behavior founded on scientific theories. There have been a few attempts, but they were abandoned. Saint-Simon in the late 1700s, Auguste Compte in the early 1800s, and Emile Durkheim in the late 1800s and early 1900s were at pains to develop a set of "positive"—that is, scientific observation- and experiment-based—codes for behavior. But this endeavor was contrary to the underlying ideal of "pure objectivity" of modern science and it was not taken up by the mainstream of the scientific community.

In the 1990s, however, both scientists and political leaders began to recognize the need for principles that would state universal norms for behavior. The Union of Concerned Scientists, an organization of leading scientists, issued a statement in 1993 signed by 1,670 scientists from 70 countries, including 102 Nobel laureates. "A new ethic is required," the statement declared, "This ethic must motivate a great movement, convincing reluctant leaders and reluctant governments and reluctant peoples themselves to effect the needed changes." The scientists noted our new responsibility for caring for the earth and warned that "a great change in our stewardship of the earth and the life on it is required if vast human misery is to be avoided and our global home on this planet is not to be irretrievably mutilated." Human beings and the natural world, they said, are on a collision course, one that could so alter the living world that it will be unable to sustain life as we know it.

In November 2003, a group of Nobel Peace laureates meeting in Rome affirmed, "Ethics in the relations between nations and in government policies is of paramount importance. Nations must treat other nations as they wish to be treated. The most powerful nations must remember that as they do, so shall others do." And in November 2004, the same group of laureates declared, "Only by reaffirming our shared ethical values— respect for human rights and fundamental freedoms—and by observing democratic principles, within and amongst countries, can terrorism be defeated. We must address the root causes of terrorism—poverty, ignorance, and injustice—rather than responding to violence with violence."

An ethics is effective in practice only if people find it meaningful and worthy of living by. This holds true for a planetary ethics. This ethics must state the absolute minimum standards for behavior that are consistent with responsible living on this planet. It is very different from, and is indeed diametrically opposed to, an ethics of might is right—the "survival of the fittest" ethics. The renowned biologist T. H. Huxley made this point over one hundred years ago:

> The practice of that which is ethically best—what we call goodness or virtue—involves a course of conduct that, in all respects, is opposed to that which leads to success in the cosmic struggle for existence. In place of ruthless self-assertion it demands self-restraint; in place of thrusting aside, or treading down, all competitors, it requires that the individual shall not merely respect, but shall help his fellows; its influence is directed, not so much to the survival of the fittest, as to the fitting of as many as possible to survive.*

Today, on a planet of nearly seven billion, living with dwindling resources and in a deteriorating environment, adopting an ethics of "fitting as many as possible to survive" is essential. The basic principle of a planetary ethics must be: *live in a way that all others can also live.*

---

*T. H. Huxley, "The Struggle for Existence in Human Society." In *Evolution and Ethics and Other Essays.* T. H. Huxley, ed. (New York and London: Appleton, 1925).

Though this principle is simple and meaningful, it is not easy to abide by. Whether consciously or not, most people subscribe to the Darwinian "survival of the fittest" ethics, or else to the ethics of classical liberalism: "live and let live." You can compete in any way you want, and you can do as you please, as long as you don't break any laws (or as long as you are not caught at it). Today this is not enough. Sometimes legally but never legitimately, the rich and the powerful marginalize their less endowed competitors, consume a disproportionate share of the resources that need to be shared by everyone, and voluntarily or inadvertently block access to vital resources for the poor and the powerless. When the aim is the survival of the fittest, the majority of the people cannot be made fit to survive.

Abiding by the principle *live in a way that all others can also live* calls for the rich to change their objectives, their lifestyle, and their way of consumption. The planet has neither the resources nor the carrying capacity for all people to drive private cars, live in separate homes, and use the myriad gadgets and appliances that go with the lifestyle of affluence. We must all live more fairly and responsibly.

A planetary ethics also calls for changes on the part of the poor and the recently rich: they, in turn, must cease to emulate the lifestyles of the already established rich. It would not be enough for Americans, Europeans, Japanese, and Australians to reduce harmful emissions and economize on energy if the Chinese and the Indians acquire Western driving and consumer habits and pollute the air, the lands, the rivers, and the seas. Whether you are rich or poor, "developed" or "developing," to live ethically you must make sure that the way you live allows all people on the planet to live.

In the fifth century B.C.E., in the *Tao Te Ching,* Lao-tzu wrote,

One's individual life serves as an example for other individuals; one's family serves as a model for other families; one's community serves as a standard for other communities; one's state serves as a measure for other states; and one's country serves as an ideal for all countries.

A planetary ethics means that each individual, each community, and each country adopts the kind of lifestyle and the kind of consumption that's a fair measure for the lifestyle and way of consumption of all individuals, communities, and countries throughout the world.

Evolve Your Consciousness

Without a global revolution in the sphere of human consciousness, nothing will change for the better. . . . And the catastrophe towards which this world is headed—the ecological, social, demographic, or general breakdown of civilization—will be unavoidable.

VÁCLAV HAVEL,
ADDRESS TO THE U.S. CONGRESS, 1991

This frequently cited pronouncement by the former President of Czechoslovakia makes a good point, but it's not a cause for despair—it's not a forecast but an "if-then" proposition. *If* a global revolution in consciousness fails to come about, *then* there will be a general breakdown of civilization. But a global consciousness revolution *could* come about. Bringing it about is a major objective of our time.

Notwithstanding the skeptical view—that people have always been selfish, greedy, and power-hungry, and always will be—history testifies that consciousness is not an unchanging fixture of our species. There have been very different kinds of societies in the past, and the people who lived in them had a consciousness very different from that which prevails today. The difference between a traditional society, an archaic society, a classical society, a medieval society, an oriental society, and modern Western society is not just a difference in wealth and technology. It's first and foremost a difference in the consciousness of its people.

Consciousness has changed in the past, and it can change again in the future. A positive change is urgent and crucial. How could people shift their values, perceptions, and behaviors unless they evolve their consciousness? How could they come up with the will to pull together to confront the threats they face in common, and elect political leaders who support projects of cooperation and solidarity? How could they create the kind of demand that ethical businesses could satisfy with fair and sustainable products and services? Without a more evolved consciousness the motivation for change would have to await the coming of crises and catastrophes—and if these have already reached the point of no return, it will be too late.

The next evolution of consciousness. Consciousness could continue to evolve in our time, and it could evolve in good time. But what would a more evolved consciousness be like? Some mystics, philosophers, spiritual leaders, and social scientists have given serious thought to this question.

The Indian sage Sri Aurobindo viewed the emergence of what he termed "superconsciousness" in at first a few and then ever more individuals as the mark of the next higher stage of human consciousness. (Superconsciousness is the kind of consciousness that already occurs in samadhi, satori, and similar states of meditation and enlightenment.) The Swiss philosopher Jean Gebser defined the next stage as the coming of four-dimensional integral consciousness, arising from the prior stages of archaic, magical, and mythical consciousness. The American mystic Richard Bucke portrayed this stage as cosmic consciousness, beyond the simple consciousness of animals and the self-consciousness of contemporary humans.

For the mystic Eckhart Tolle consciousness is part of the universe: the essential part. It's the intelligence, the organizing principle behind the cosmic arising of form, which is the basic evolutionary process. Through evolution, Tolle says, consciousness has been preparing forms for millions of years, so it could express itself through them. Today consciousness is ready to create form without losing itself in it—it can remain aware of itself, even while creating and experiencing form. Thus the next stage in the evolution of human consciousness is the state of awakening—the consciousness of mastering the art of "awakened doing."

Social scientists Chris Cowan and Don Beck elaborate a colorful scheme called spiral dynamics. According to this concept human consciousness evolved from the strategic "orange" stage, which is materialistic, consumerist, and success-, image-, status-, and growth-oriented to the consensual "green" stage of egalitarianism and orientation toward feelings, authenticity, sharing, caring, and community, and is now shifting to the ecological "yellow" stage where it's focused on natural systems, self-organization, multiple realities, and knowledge. In the future it could reach the holistic "turquoise" stage of collective individualism, cosmic spirituality, and Earth changes.

Philosopher Ken Wilber describes a six-level evolution that has led from the physical consciousness of the nonliving world through the biological consciousness of animals to the mental consciousness of present-day humanity, which could in turn issue in a subtle consciousness that is archetypal, transindividual, and intuitive. This evolution would lead to "causal consciousness" and culminate in the ultimate consciousness Wilber calls Consciousness as Such.

Spiritual traditions also speak of the coming of a new consciousness. The prediction of the Mayan elders is remarkably aligned with the above insights. The Mayans say that the coming era will be an era when the ether, the long-neglected fifth element of the universe, will become dominant. "Whereas the four traditional elements [air, water, fire, and earth] . . . have dominated various epochs in the past," spokesperson and Mayan high-priest Carlos Barrios says, "there will be a fifth element to reckon with in the time of the Fifth Sun: ether." Ether is a medium, he pointed out, it permeates all space and transmits waves of energy in a wide range of frequencies. An important task at this time is "to learn to sense or see the energy of everyone and everything: people, plants, animals. This becomes increasingly important as we draw close to the World of the Fifth Sun, for it is associated with the element ether—the realm where energy lives and weaves" (see www.SacredRoad.org).

Coincidentally, but perhaps not accidentally, physicists are discovering that the ether was not rightly discarded one hundred years ago, when experiments failed to detect the friction it was predicted to cause in the rotation of Earth. The place of the ether is not replaced by empty space, the so-called vacuum. What physicists now call the quantum vacuum is not empty space: according to grand-unified theories it's the unified field, the womb of all the fields and forces of nature. It contains a staggering concentration of energy, and carries and transmits information.

In Sanskrit and Hindu philosophy the ether was considered the most fundamental of the five elements; the one out of which the others arose. The ether was known as *Akasha,* the element that *connects* all things—this is what we now call the Akashic Field—and that conserves the *memory* of all things—known as the Akashic Records. Today, as an energy- and information-field rediscovered in the universe, the ether regains much of the status it enjoyed five thousand years ago.*

A consciousness that recognizes our connections to each other and to the cosmos is an Akashic consciousness—a consciousness of connectedness and memory. It conveys a sense of belonging, ultimately, of oneness. It's a wellspring of empathy with nature and solidarity among people, the kind of consciousness foreseen by mystics and philosophers from Aurobindo to Wilber, predicted by the Mayans, and now supported by discoveries at the leading edge of the sciences.

The evolution of a new consciousness is not utopian: it's already under way. There is considerable evidence gathered by psychologists, psychiatrists, experimental parapsychologists, sociologists, and even brain researchers regarding the emergence of a form of mind and consciousness variously called integral consciousness, extended mind, nonlocal consciousness, holotropic mind, infinite mind, and boundless mind.†

---

*For more on the Akashic field, see my *Science and the Akashic Field* (Rochester, Vt.: Inner Traditions, 2007).

†I have explored the nature and possibilities of Akashic consciousness in *The Akashic Experience: Science and the Cosmic Memory Field* (Rochester, Vt.: Inner Traditions, 2009).

Some aspects of this advanced consciousness have been measured. English electrical engineer C. Maxwell Cade tested the electrical activity of the brains of more than three thousand individuals and described his results in his 1979 book *Biofeedback and the Development of Higher States of Awareness*. His subjects included accomplished natural healers, who presumably possess an advanced transpersonal consciousness. Cade found a typical pattern of EEG (electroencephalograph) waves appearing in the healers' brains while they are engaged in healing. The pattern he found consists of a moderate amount of beta and theta waves, a wide band of alpha waves, and no waves in the delta region (the known frequencies are beta, with a range between 8 and 30 Hertz [frequency per second]; alpha, ranging from 8 to 13 Hz; theta, between 4 and 7 Hz; and delta, in the range of 0.5 to 4 Hz).

This finding is significant, since the typical EEG pattern of modern people is almost entirely in the beta range. Alpha occurs only in meditation and restful states, theta in half-awake or dreaming sleep states, and delta in profound dreamless sleep. Except for the presence of beta waves (which signify brain activity in the context of everyday awareness), the pattern Cade found in the healers is the same that appears in samadhi, satori, lucid dreaming, and various mystical states. In some cases this pattern has proven remarkably stable, persisting even in their everyday life. It appears to have become their natural state of consciousness.

Furthering the evolution of our consciousness. If a breakdown is to be averted, the evolution of human consciousness must be rapid and widespread. It's up to each person to bring it about. Consciousness evolves in the individual, and spreads from one individual to another. Its growth starts with you, the same as with me. *But how do you evolve your consciousness?*

This is an eminently practical question. There are many ways to evolve your consciousness, some calling for conditions and events that are beyond your control, while others are entirely within the domain of your ability and will.

A serious illness or accident, especially if it involves returning from the portals of death, tends to be a life-transforming experience. It often brings a marked change in consciousness. People come back to life with a very different spirit: they possess inner peace and empathy for others, they have reverence for nature, and a fresh appreciation of the wonder of existence. Astronauts who had the privilege of looking at Earth from outer space come back with a similarly changed consciousness. Apollo 14 captain Edgar Mitchell reported a veritable epiphany he experienced as he viewed our blue-green home planet swimming in cosmic space. His values and beliefs changed fundamentally; his life took on a radically new turn.

But you don't need to be shot into space or nearly die to evolve your consciousness. There are simpler ways.

To begin with, you need to have meaningful contact with your body. Being constantly occupied and preoccupied with tasks and worries, hopes and fears, we all have lost contact with our body—we use it as we use our car: giving it commands to take us where we want to go. Contacting your body calls for simple exercises in bodily movement, for conscious breathing, and for concentration on what is going on in one part of your body after another. The techniques are not difficult, and they are taught by many psychologists and spiritual teachers.

You also need to pay more attention to your feelings and emotions. If you are like most people, you have not lost contact with your emotions, but the emotions you feel are mostly of the wrong kind. Anger, hate, fear, anxiety, suspicion, jealousy, contempt, and indifference dominate the tenor of contemporary life. They result from negative life experiences. Even childhood education is based on the threat of punishment and fear of failure. Positive emotions of love and caring are sacrificed to the pressure of work and the struggle to secure one's livelihood. Yet in your relations with family, friends, collaborators, and fellow members of the community you could enjoy positive emotions of love and caring. The essence of love is a deep feeling of connection, whether it's love for a child, for a parent, for a lover, or for a friend.

Another way to advance your consciousness is to take time to contemplate nature. The tranquillity of a sunny meadow or a calm lake, the beauty of a sunset, the majesty of a mountain, or the awesome force of the stormy sea convey ineluctable feelings of oneness and belonging, the hallmarks of an evolved consciousness.

Feelings of oneness, love, and belonging are also generated in the experience of deep prayer. In his seminal work *The Varieties of Religious Experience,* William James noted that in a genuine religious experience there is a sense of union with something or someone larger than ourselves. We become aware that we are like islands in the sea: separate on the surface, but connected in the deep.

Deep meditation conveys the same insight. As you still the mind, free it from the fears, fantasies, and concerns about what will or may happen, consciousness becomes purified: it remains present, but it is calm and still. This is the meaning of *samadhi,* the state of "still mind." Physicist-philosopher Peter Russell noted that when you allow the mind to sink into the silence of pure consciousness the qualities that usually distinguish one self from another are no longer there—all markers of individuality are gone. You become aware that you are the light of consciousness, the same light that shines within all beings. You become one with all beings.*
Indeed, the original meaning of the word *yoga* in Sanskrit is "union."

Eckhart Tolle speaks about the "power of now." This is a state where you experience your own self, without thinking, without emotions, a state of pure awareness. In this state, personal powerment consultant Faye Mandell tells us in her 2003 book *Self-Powerment: Towards a New Way of Living,* you know intuitively that you are connected to everyone. Your entire experience of self changes. You detach yourself from the personal image of yourself and experience the universal energy that integrates all of us. You become grounded and centered, and experience your self without selfishness.

There are many ways to evolve your consciousness, and the right ways always lead to the experience of connectedness with others and

---

*Peter Russell, *Waking Up in Time: Finding Inner Peace in Times of Accelerating Change,* tenth anniversary edition (San Rafael, Calif.: Origin Press, 2009).

with nature—to the experience of oneness. In this state of consciousness you are "in syntony with"—that is, finely tuned to—the world around you. According to Alexander and Kathia Laszlo, who devote their life to seeking syntony, this can be achieved on the personal, interpersonal, and transpersonal level.

> The first level of syntony is *personal* syntony. It requires centering, quieting the mind, listening with every cell of your being so as to cultivate intuition and compassion, insight that matches outsight with a willingness to explore and follow your deepest calling.
>
> The second level of syntony is *interpersonal* syntony. It involves dialogue and collaboration with people around you, coming together to learn with and from each other and engaging in joint action with empathy, considerateness, openness, and joy.
>
> The third level of syntony is *transpersonal* syntony. Finding syntony on this level calls for "communing"; for listening to the messages of all beings whether they are waterfalls, animals, mountains, or galaxies, and feeling and acknowledging your oneness with all things in the cosmos.

"As an organizing force in societal evolution," Alexander and Kathia Laszlo write, "syntony involves an embodiment and manifestation of conscious evolution: when conscious intention is aligned with evolutionary purpose, we can foster and design evolutionarily consonant pathways of human development in partnership with Earth. The effort to cultivate these dynamics constitutes a syntony quest."*

---

*Alexander Laszlo and Kathia Castro Laszlo, "Syntony and Flow: The Artscience of Conscious Emergence," in *The View: Mind Over Matter, Heart Over Mind—The Vital Message 2012,* David Patrick, ed. (London: Polair Publishing, 2009). Also see syntonyquest.org.

FIVE

Action!

CREATING CONSCIOUS CHANGE

How can objectives of conscious change be adopted in practice? And how can they be achieved in time to bring about a WorldShift, a shift from the edge of breakdown to a path that can lead to breakthrough, before it becomes too late?

Conscious change objectives *can* be implemented: they are intrinsically feasible. Implementing them calls for conscious, purposive action by you and by me, and by every responsible human being wherever they live on this planet.

CIVIC ACTIVISM

Let us start with what you, as a responsible person, can and should do in the civic sphere. Civic activism means interacting with the people and the institutions that govern your state or community. Doing so is important because, although states have lost much of their sovereignty and governments have lost much of their power, the role of the public sector remains decisive. There are problems of peace, security, and public well-being that only a government can address, and only an enlightened government can address well. It's essential, therefore, that you and others around you nominate and elect political leaders who genuinely serve the public interest, and that you interact with the elected leaders to prompt them to serve it effectively.

The most basic thing you must ask of your government is that it recognizes and respects the basic values of life in the community. This means concern with people's access to adequate supplies of food, water, shelter, and clothing. It means concern with stable relations in family and community. It means regard for the social and cultural identity and economic equity of all the people who share the community. And it means concern with the availability of useful and remunerated employment for everyone who can and is willing to work.

Beyond these basic objectives, you can ask that your government adopts objectives that respond to requirements that are specific to, and crucial for, our time of transformation. Pursuing such objectives calls for the government to

- use criteria of sustainability in the allocation of natural and financial resources;
- ensure access to energy, water, and land to all people at acceptable cost;
- increase the quality and capacity of the public transportation system;
- promote cooperation with business companies in the use of alternative energy and resource technologies;
- channel funds and provide incentives for the revitalization of derelict areas;
- employ strict criteria of sustainability and energy and resource efficiency in licensing new construction;
- bring social and ecological responsibility considerations into the system of public education; and
- make nature accessible to people while safeguarding ecological balances and wilderness areas.

You can insist that your government comes up with the funds to implement policies that genuinely serve the public interest. Even in a condition of economic and financial crisis, this is not an unreasonable demand: there is much unused or badly used wealth in today's world. Even a fraction of the enormous amounts of money circling the globe could fund projects that serve human well-being. As it is, less than 5 percent of the flow of money over the globe is for real goods and services. Of the remaining 95 percent a small portion is devoted to legitimate investments, but the rest—over a trillion dollars a day—is used for speculation: to make more money with the money. And even of legitimate investments, large amounts are chan-neled to short-term projects of questionable human benefit.

In industrial countries over 250 billion dollars a year could be redi-rected from coal and oil-based energy production to carbon-free technolo-gies. A fund of about 300 billion dollars a year could transfer clean energy technologies to the developing countries. Even a small part of the over $1.2 trillion dollars that the world's governments spend annually on arms, wars, and the military could implement urgently needed projects. According to United Nations' estimates, an annual expenditure of $125–150 billion dol-lars would feed the world's hungry, ensure an adequate supply of clean water, overcome abject poverty, treat AIDS, eliminate diseases such as malaria, and restore vital balances in local and continental ecosystems.

The March 2009 decision of the G20 group of industrialized coun-tries to pump trillions of dollars into the global economy testifies that there is money in the hands of governments if they have the will to use it. But the money they have available needs to be used for wise, long-term peace- and sustainability-oriented projects and not just for short-term economic revitalization. The determined insistence of people like you in the great democracies of the world is a basic precondition for accomplishing these goals.

BUSINESS ACTIVISM

Business companies, we have noted, can no longer reasonably view them-selves as "private sector" organizations divorced from the concerns of life and well-being in society; they need to become dedicated to furthering the public good. Grassroots activism can promote this shift. You don't have to be an executive of a business or even an employee to act effectively: it's enough to be a consumer, a client, a member of the host community, and perhaps the owner of a few shares in the company.

A simple and effective way to further the shift in the business world is discriminating consumption. This indirect yet effective form of activism calls for patronizing firms that embrace a commitment to the public good and keeping away from those that remain committed only to their own immediate self-interests.

There are more direct ways you can influence the actions of a company. If you have the means to buy a few shares, you can join a shareholder association and raise issues of social and ecological responsibility in shareholder assemblies. Through such associations even small shareholders can insist on transparency in corporate reporting, and ask that management discloses how the company treats its employees and business partners and relates to its host communities and the environment.

If you have the means, you can practice "social investing." Social investing started when religious investors, such as the Quakers, began to avoid investing in so-called sin industries: alcohol, tobacco, gaming, pornography, and military suppliers. In the 1980s many investors boycotted companies that supported the apartheid regime in South Africa, and by the early 1990s large social mutual-fund companies began to give priority to companies that met rigorous social and ecological criteria.

Social investing is not charity, and it doesn't call for financial sacrifice. There are ethical investment funds that carefully screen the shares they acquire, and in the medium- to long-term investors often do better using them than by investing indiscriminately for short-term profit.

As a shareholder, client, consumer, and member of the host community, you can interact with business companies on a variety of issues. Among other things, you can insist that the company

- accurately and honestly represents to the public the long-term benefits and costs of its products and services, reporting on their safety, social consequences, environmental toxicity, reusability and recyclability;

- actively seeks to reduce pollution and environmental damage and minimizes waste in its production processes and throughout the chain of supply and distribution;
- consults employees and collaborators when formulating its goals and objectives;
- gives preference to responsible companies as partners and associates, and refuses to do business with companies that behave unfairly toward their employees, customers, and host communities, and degrade the environment;
- takes an active interest in the lives of its employees, discovering their concerns, understanding their needs, and contributing to their personal development; and
- takes a similarly active interest in its host communities, allowing and even encouraging employees to devote part of their time to social work, and the improvement of the local environment.

Insisting that companies embrace the social objective is not asking them to act against their own interest. In the business world doing good and doing well do not exclude each other. Quite the contrary. Companies that do good can count on a higher level of public confidence and greater customer and client loyalty. They are likely to be more successful than narrowly self-centered enterprises, having a better chance to survive the fluctuations and crises that rock the world of business.

MEDIA ACTIVISM

The flow of information in today's world is enormous, yet it's largely irrelevant to the issues that decide the quality, the security, and the prospects of people's lives.

For the commercial media short-term market research decides what information reaches the public: people get what media executives believe they are interested in. Believing that the public is interested in few things other than war, crime, sports, the stock market, and the views and doings of celebrities, media executives concentrate on sensational items with "news value" and "human interest." But an exclusive diet of such items doesn't equip you and your family, friends, and neighbors with the insight and information needed to decide the issues that concern our life and our future.

The media could operate more responsibly, and you have the power to make it do so. You can select television and radio programs, Internet sites, newspapers, magazines, and books that provide pertinent and constructive news, rather than merely sensational items. A number of alternative papers and magazines have now become available, published by socially conscious not-for-profit organizations. They have a growing readership. If you and others around you show a real interest in getting relevant news and information, these media will continue to grow, and there will be a page or column with responsible content also in mainstream papers and magazines, and more such programs on commercial radio and television.

Requiring the information and communications media to become relevant rather than merely sensational is not asking that it become charitable: being relevant to people's concerns is in the media's own interest. Responsible media executives could undertake what the *Harvard Business Review* called "expeditionary marketing": creating the kind of demand that's in the public's best interest, rather than responding to the lowest common denominator in the current demand.

Public demand is shifting already. The demand is growing for news and information about cooperative communities, ethical movements, ways of sustainable living, alternative energy sources, efficient, environmentally friendly technologies, and non-polluting products. Furthering this trend is good for the media's clients and customers, and good for the media itself.

PERSONAL ACTIVISM

Some aspects of your private life have become public business. What you do affects others, and either furthers a transformation toward a better world, or blocks it. Whether you know it or not, you are either an agent of change, or the rear guard of the old and obsolete culture.

As a conscious agent of change, you must live responsibly. This means:

- living in a way that satisfies your needs without detracting from the opportunity of other people to satisfy theirs;
- living in a way that respects the right to life and development of all people, wherever they live, and whatever their ethnic origin, sex, citizenship, and belief system;
- living in a way that safeguards the right to life and a healthy environment of all things that live and grow on Earth;
- pursuing happiness, freedom, and personal fulfillment in consideration of the similar pursuits of your fellows in your community, country, and culture, and in the global community of all peoples, countries, and cultures;
- choosing a work or profession and committing your time and talents to an activity that is useful and beneficial to your community and doesn't harm other people, other communities, or nature;
- doing your best to help children and young people to discover their own responsible ways of thinking and acting; and
- joining forces with like-minded people to preserve or restore the integrity of the environment so it can generate and regenerate the resources essential for human life and well-being.

As a conscious agent of change, you need to adopt a responsible lifestyle. This means increasing the sustainability of your household by reducing its carbon footprint, saving energy and natural resources, and minimizing waste. The ways you can do this are widely known. You can save energy in the home by using energy-saving lights and turning them off when not needed; you can insulate your lodgings against heat and cold instead of turning on the air conditioner in summer and jacking up the furnace in winter; you can save water by installing simple devices that produce a stream of water at adequate pressure at a low rate of consumption; and you can save energy and cut down on carbon emissions by using public transportation, or by walking or riding a bicycle—among other things.

Discriminating consumption is another facet of personal activism. This means selecting clothes that express your personality and cultural values rather than prestige labels; choosing home furnishings made of long-lasting natural materials that make for warmth and sociability instead of ostentatious items that show how much you can afford; and using household appliances, home and garden devices and similar goods that are made to last, are locally produced, and are energy efficient.

Choosing foods for your table that are healthy, organic, locally or regionally grown, and don't require vast amounts of energy and water to produce is particularly important. The diet of the average household in America accounts for the emission of 2.8 tons of carbon dioxide a year, compared with 2.2 tons emitted in driving. Replacing the consumption of red meat, pork, dairy products, and processed snacks with vegetables offers significant carbon savings and is at the same time healthier: grains, fruits, and vegetables require about 2 calories of fossil-fuel energy to yield 1 calorie of food energy, whereas the ratio for beef can be as high as 80 to 1.

Being an agent of conscious change is not difficult, and it doesn't call for sacrifices. On the contrary, it brings many benefits. You become a better friend and neighbor, live more healthily, and have the satisfaction of knowing that you are doing your best to ensure that you and your children live to see a world of sustainability and peace, instead of a world of mounting stress and violence.

Harry Truman once said, "The buck stops here," meaning the desk in the Oval Office. Today the buck is more democratic: it stops with every one of us. We live in critical times and we, and only we, can make the crucial difference. The factor that decides our future is not wealth and power, and not the control of territory and technology. It's how we think and act.

Part Three

BREAKTHROUGH 2032

A Thought Experiment

We started this handbook of conscious change saying that today's world is unstable, unfair, and prone to violence. We end it showing that this condition is not endemic. Our world could also be stable, fair, and peaceful.

Can a stable, fair, and peaceful world be created in practice—and in time? Can people change their thinking and behavior from today's selfish and self-centered material-gain and power-oriented ways to cooperative and sustainable ways? The answer is yes: *through conscious change by a critical mass*. But can conscious change be embraced by a critical mass before current trends and problems become intractable? The answer is still yes: *by accelerating the spread of the consciousness that's already emerging at society's creative edge*. The rapid spread of an evolved consciousness is a basic precondition of moving toward an effective and timely WorldShift.

If a WorldShift occurred by or shortly after the end of 2012, a more evolved consciousness could surface by 2032, when the young generation then growing into positions of responsibility would have had two decades to reflect and to learn.

We can illustrate this possibility with a thought experiment. We go into fast-forward and project ourselves to 2032. A stable, fair, and peaceful world has been created. The question we pose is, what kind of values and consciousness hallmark people's thinking and acting in that world, and what kind of structures, institutions, and technologies have they adopted to achieve their aspirations?

SIX

Reports from a Peaceful and Sustainable World

We are in the year 2032. We query two young people about how most people think and how they pursue their aims and goals. A young woman, a schoolteacher, tells us about the way a critical mass of the people think: their values and consciousness. And a young man, a community counselor, recounts how the 2032 world is structured, what kind of institutions it has, and what technologies it adopts. They then join together to report on how their world grows and develops, without conflict and violence and irreversible damage to the environment.

THE WAY WE THINK:
THE REPORT OF THE YOUNG SCHOOLTEACHER

There are many countries and cultures in our world, some large, other smaller; some industrialized, others predominantly rural. There are a few big cities, but most people prefer to live in medium-size cities and towns, or in the countryside. On the whole, abject poverty has been eliminated: everyone's right to food, housing, education, and socially useful remunerated work is recognized and respected. Because life is less stressful, diversity can flourish without being constrained by everyday survival concerns. North Americans and Latin Americans, Japanese, Chinese, Indians, and Asians, the same as Europeans, Africans, Australians, and Polynesians, can express their values and safeguard or develop their traditions.

People live longer and healthier lives, but the population is not increasing. Longer life spans are offset by smaller families, as people realize that it's irresponsible to produce more children than they can care for. This has obvious advantages. With a modest family size people are able to take better care of their children, ensuring that they grow into healthy individuals, with sufficient education to live peacefully and sustainably, in harmony with society and with nature.

Our Values and Our Consciousness

The thinking that orients our life is not the same kind of thinking that prevailed prior to 2012. Our thinking has changed, but not by rules and legislation and fears of punishment, but by adopting new values and evolving our consciousness.

Our values and consciousness are very different from those that dominated the world when I was a young girl, in the first decade of the twenty-first century. They came about in the aftermath of the WorldShift of 2012, when we realized that we cannot have peace and sustainability on a small and highly interacting and interdependent planet unless we cooperate—and that we cannot muster the will to cooperate unless we share basic values and aspirations.

The shared fundamental value of our world is inner rather than outer growth—the growth of social, intellectual, and emotional life. We hold those in esteem who have integrity and sound ethics, who contribute to a better life for family, community, and humanity, who safeguard the balances and processes of nature, and who seek to discover and convey deeper insight into the meaning of life.

Personal achievement used to be measured by the amount of money a person had and the level of power he or she wielded. This is not what we think any longer. The possession of money is not the measure of personal achievement, and material goods are not its outer sign. The symbol of social status is not a bigger home and car, and the pinnacle is not the yacht and the private jet. Being rich is defined not by *having*, but by *being*.

Possessing material goods beyond what's needed to ensure a decent quality of life is not a sign of achievement; on the contrary, it's an indication of backward thinking. Real wealth lies not in the possession of money, but in living a fulfilling life, with loving families, healthy and happy children, a caring community, and a healthy environment. Living well means living wisely, comfortably, even luxuriously, yet comfort and luxury are not measured by the quantity of the goods we own and control, but by the quality of our life and our lived experience. These values and aspirations are present in the way we educate our children and young adults. Our systems of education have changed. While the specific features vary from community to community and region to region, the curricula of schools and universities have elements in common. Their basic mission is to educate a generation of locally active and globally thinking planetary citizens, able to live a healthy, productive, and responsible life, in harmony with each other and with nature.

Learning is not segmented into individual compartments and disciplines but uses a holistic, systemic framework. It encompasses four major topics: ecological sustainability, public health and welfare, cross-cultural understanding, and local and global economic well-being through the fair and sustainable enjoyment of the planet's resources. The new consciousness is not taught: it emerges spontaneously, and is often more evolved in our students than in us, their teachers.

The new values inspire ambitions and aspirations that embrace all aspects of our life.

- We aspire to democracy, yet for more than a formal democracy; we aim for a participatory society, where all people have a voice in making the decisions that affect their life and future.
- We aspire to fairness toward all people, but we aspire to more

than mere fairness; we aspire to goodwill among peoples, nations, and cultures.

- We aim for the rule of law, but for more than the means to enforce laws; we aspire to genuine respect for laws that protect the rights and security of individuals and communities.

- We aspire to achievement in the material sphere, but for more than the accumulation of material goods; we aspire to being in a position to help create and sustain economic conditions where people can enjoy a decent material standard of living.

- We aim at social development, but at more than developed social structures and infrastructures; we wish to build a society that ensures a high quality of life for its people.

- We aspire to tolerance for differences among people and cultures, but we aspire to more than passive tolerance; we seek the active pursuit of mutual interests, building on the shared values that underlie our differences.

- We aim for freedom from oppression, hunger, and misery, but for more than that; we aim for the freedom for everyone to construct his or her own existence, through activities and lifestyles of his or her responsible choice.

- Last but not least, we aim for humanism in our actions and decisions, but not for an abstract kind of humanism; our humanism is to ensure that we ourselves live in a way that allows all people on this planet to live.

Our values and aspirations derive from a common source: our consciousness. A changed consciousness has been spreading among us in the last two decades. It had precursors in the past, but this kind of consciousness was considered the exclusive province of healers, religious leaders, spiritual masters, artists, and philosophers. It's no longer that.

Before the WorldShift my parents thought of empathy, intuition, and spontaneous insight as esoteric or "New Age." They were surprised when more and more children were born with a capacity for them. These children were at first called "indigo children" but, as time passed, the new mental capacities became the norm rather than the exception and terms such as *indigo* and *New Age* were left behind.

The new consciousness is in stark contrast to the narrow ego-bound and brain-limited consciousness of the industrial age—it's a transpersonal consciousness, open to people and the world. It enables us to connect with the element that the ancients called ether, and that scientists rediscovered as the Akashic Field, a fundamental aspect of the unified field that pervades cosmic space. Because this energy and information field links all things with all other things, connecting with it means overcoming the narrow bounds of the consciousness that hallmarked the previous age.

Thanks to our transpersonal "Akashic" consciousness, we sense and feel our connections with things large or small, near or far, in a subtle and intuitive yet meaningful way. This inspires solidarity among us, and empathy with nature. It helps us create syntony in the world. Thanks to our new values, new consciousness, and new ways of pursuing and disseminating knowledge, a critical mass among us has reached deeper insights.

- We know, and feel with every cell of our body, that we are not separate entities pursuing independent destinies but interconnected elements of a reality that is larger than each of us and embraces and makes whole all of us.
- We know and feel that all seven-and-a-half billion of us are inhabitants of the earth, with an equal right to enjoy its life-supporting resources and an equal responsibility to safeguard them for coming generations.
- We recognize that nature is not a mechanism to be engineered and exploited, but a living system that brought us into being, that nourishes us, and—given our awesome powers of exploitation and destruction—is now entrusted to our care.
- We are convinced that it is immoral for any of us to live in a way

that detracts from the chances of the rest of us to achieve a life of basic well-being and human dignity.

- We believe that the universal rights adopted by our forebears in the twentieth century—the right to freedom of expression, freedom to elect our leaders, and freedom from torture and other arbitrary constraints on personal liberty, as well as the right to food, shelter, education, and employment—apply to everyone in the global community, and deserve to be respected above and beyond considerations of personal, ethnic, and national self-interest.
- We have learned that the way to solve our problems and disagreements is not by attacking each other, but by cooperating in ways that serve our common interests.
- And we have learned that the best way to overcome discord and contention is by opening ourselves up to our adversaries, and realizing that, in the context of humankind's fundamental oneness, our differences are petty and superficial.

THE WAY WE ACT: *THE REPORT OF THE YOUNG COMMUNITY COUNSELOR*

The fears that dominated the first decade of this century—fears of terrorism, armed conflict, economic breakdown, famine, ecological collapse, and invasion by destitute migrants—are behind us. Stability is the hallmark of this world. This is not the rigid stability imposed by a powerful authority, but the stability of a sustainable network of self-reliant but cooperative communities, states, nations, and federations of nations.

The states, nations, and federations are free to choose their preferred social structures and economic systems. This doesn't create irremediable conflict. We have not fragmented the human community into isolated units pursuing separate goals without regard for the common good. The many diverse nations and cultures are united by common values and aspirations, centered on creating a world where all people can live safely and peacefully, without destroying each other and their life-sustaining environment.

Our Politics

The twentieth century's system of self-centered nation-states has been transformed into a transnational system, organized as a series of administrative and decision-making forums, where each forum has its own sphere of competence. The forums have considerable autonomy; those at the lower levels are not subordinated to forums at the higher rungs. Our political system is not a hierarchy, but a multilevel structure of distributed decision making. It seeks to overcome the contradiction between global coordination and local, regional, and national autonomy.

Multiple links of communication and cooperation crisscross this interlinked social, political, and economic system. Individuals jointly shape and develop their local communities. These communities participate in a wider network of cooperation that includes, but does not stop at, the level of nations. Nation-states in turn are part of continental or subcontinental social and economic federations.

Global forums decide questions of peace and security, the protection of the environment, information and communication, and international finance. The principal forum at this level is the United Peoples' Organization, the body that succeeded the United Nations. The UPO observes, as do all other decision-making organs, the well-known but previously seldom-respected "principle of subsidiarity": decisions are to be made at the lowest level at which they can be effective. The global level of the UPO, the world's highest level of decision making, is at the same time the lowest level at which peace and security can be effectively safeguarded, the world environment can be cared for, and the flow of money, technology, and information across the continents can be regulated. All other issues of public policy are delegated to local communities, to national states, and to the federations of national states.

The UPO's political members are regional federations of national states. These embracing social and economic bodies grew out of the experience of the European Union. They now include the European Union, the North American Union, the Latin American Union, the North-African Middle-Eastern Union, the Sub-Saharan African Union, the Central Asian Union, the South and Southeast Asian Union, and the Australian-Nippon-Pacific Union.

The federations jointly constitute the Peacekeeping Council of the UPO, holding a mandate similar to that of the former Security Council, but without the two-tier structure where some members are permanent and others not, and some have veto-power and others do not. The Peacekeeping Council commands the sole significant military force in the world: the United Peacekeeping Force. The UPF, staffed by contingents from the federations, undertakes peacekeeping missions on the request of the UPO in consultation with the federation where the mission is to be carried out.

The United Peoples' Organization is not uniquely a political organization: it has members from civil society and from business. Civil society members include federations of the many thousands—according to some estimates over a million—non-governmental organizations active in the social, economic, and environmental domains. Through their representation in the UPO, NGOs have become an integral part of the deliberations that decide the sustainability, the fairness, and the responsible development of our communities.

Similar to the UPO's civil-society membership, the corporate membership is made up of federations of businesses in the major branches of industry. Through specialized agencies in finance, manufacturing, commerce, and labor, inherited from the United Nations and reformed in the light of the Organization's enlarged mandate, the UPO connects its member business federations with the representatives of the communities in which they operate. It helps managers establish good community relations, create mutually agreed upon codes of conduct, and reach mutually beneficial agreements on technology transfer, trade, employment, finance, and the protection of the environment.

Global-level coordination is a precondition of successfully restoring the viability of the environment, reestablishing natural balances in the composition of air, water, and soil, and preserving the integrity of the biosphere's regenerative cycles. In view of this requirement, the World Environment Organization has been created as a global-level partner of the United People's Organization, with the mandate to coordinate the environmental programs of the regional federations. It works closely with the political, the business, and the civil-society members of the UPO.

The continental and subcontinental level is effective for coordinating decision-making at the next subsidiary level: the level of national states. The regional federations provide a forum for the representatives of their member nations to discuss their concerns, explore areas of mutual interest, and coordinate their political goals and social-economic practices.

The tasks and responsibilities of national states have not changed significantly. National governments remain the principal arbiters of their country's economic and social objectives. The states maintain a national treasury, a national judicial system, police force, and health system. But national institutions do not operate under the premise of absolute sovereignty. Domestically they are integrated with the administrations of cities and rural areas, and internationally with the states that share a given federation.

The local level of coordination and decision making serves cities, towns, and villages. At this community level direct democracy is the rule: the representatives of the people respond directly to the people. The customary mechanism is the town hall meeting, held face-to-face whenever and wherever possible, and electronically when distance or cost prevents a significant number of people from participating in person.

Our Economics

Nation-states seek to perfect their own economics by finding the right balance between market forces, the valuation of natural assets, the environment, and social welfare. Some states provide a guaranteed income for their people to cover their basic needs. Previously unpaid work, such as maintaining a household, caring for others and for the environment, and growing one's own food, is recognized to be socially and economically useful. In this way people in the majority of today's nation-states no longer have to struggle and compete merely to obtain food, shelter, and basic education. They can choose their work, vocation, or profession free of the need to struggle for physical survival.

Sustainability in our economies is not a utopian goal; we have shifted the basic objective of economic activity. Whatever the differences in the economic systems adopted by nation-states, their ultimate goal is shared. It's not to accumulate more and more riches, but to ensure sustainability for their people and communities.

A major shift concerns our use of natural resources. We try to live on "natural income" rather than "natural capital." Natural capital consists of the accumulated riches of the earth, used and discarded, as in the burning of fossil fuels. As we discovered prior to 2012, when such capital is depleted, the economic system based on it goes bankrupt—it's unsustainable. Natural income, on the other hand, embraces all natural resources that can be renewed, replenished, or recycled. Such use of resources can be prolonged indefinitely; the economy based on it is sustainable.

In regard to human resources, the goal is not to increase the productivity of labor, making do with as few humans as possible. Managers don't try to squeeze the maximum out of every worker and every kilogram of material and kilowatt of energy. Instead, they endeavor to design production processes that provide livelihood for the local workforce, make use of the least amount of nonrenewable materials in the products, and increase the share of renewable and recyclable materials and of information.

Thanks to efficient resource use, less waste and simpler lifestyles, the size of our ecological footprint is approaching the sustainable Earthshare. The footprint of cities has dropped as well, although they remain dependent on the natural resources of their surroundings.

Business is no longer a separate "private" sector; companies have become a functional part of their host communities. As some corporate leaders did in the first part of the twentieth century, managers consider themselves builders of society. They endeavor to overcome the tension between efficiency, profitability, and dynamism on the one hand, and solidarity, equity, and sustainability on the other. They select the products and services they bring to the market in consultation with their clients and customers, as well as their employees and partners.

Production and marketing decisions are made not only with an eye toward success in the marketplace, but also in view of minimizing impact on the environment, achieving employee satisfaction, and maximizing human and social usefulness. The principal aspiration is not just to increase value for the shareholders by exploiting all the available resources, but to assume social and ecological responsibility for the stakeholders.

The valuation and use of money has changed. States, businesses, and people use money to conduct their affairs, but they don't value it for its own sake. Money is an instrument to achieve what they want to achieve, and no longer an end in itself.

Thanks to the way we use money, the built-in instability of the world's monetary system has been overcome. The severe financial crisis in the fall of 2008 marked the end of the neo-liberal belief in constant growth based on an exponentially growing return on investment. When the crisis of the financial sector sparked a recession, many attempts were made to reform the global economy, including its monetary system. Following a decade of experimentation, they resulted in the creation of a diversified system of local, regional, national, and global currencies, using the trading function of money rather than its ability to create wealth for its possessors.

In the system that came online in the 2020, U.S. dollars, Euros, and baskets of national currencies are not used as if they are global money. The UPO issues a world currency, the Gaia, on the basis of the human, and not the financial, capital of a federation; the size of its population and not the size of its treasury. The federations have their own currency, which they spend into circulation and take back through taxes. These currencies are used for trade between the federations' member states; the Gaia is reserved for interfederation transfers.

There is a whole array of complementary local and regional currencies in use for health services, education, social welfare, and culture. They function through the Internet or make use of leading-edge technologies, such as chip cards for electronic payments. Through the use of such currencies, even the least developed economies have overcome marginalization and dependence on industrialized economies and global financial flows.

The diversified yet interlinked monetary system, based on the trading function of money rather than on its ability to earn interest, ensures that global wealth is not redistributed to a progressively smaller minority. It does away with profit-motivated speculative bubbles and the inflation of national currencies. It proves to be an effective mechanism for buffering crises and instabilities in the global economy.

Our Technologies

We believe that technology, like money, is not to be valued for its own sake but for what it can do for people and society. First and foremost, it's a tool to ensure well-being and a better quality of life.

For the most part, the technologies we use are more advanced than those of the early twenty-first century, but not all the technologies that are available are put to use. Social utility and environmental friendliness are major factors in our technology choices. Technologies are to produce what's needed and beneficial without creating unwelcome side effects. Zero emission and zero impact are the ultimate goals.

A major change in the use of technology occurred in the aftermath of the deepening conflicts and escalating wars of the early part of the century. By the end of 2012 the governments of the world realized that there were no reliable measures to prevent technologies intended for defense being used for aggression, with disastrous consequences. Since in the subsequent years it was not feasible to eliminate weapons of mass destruction from the arsenal of all states, the members of the United Peoples Organization decreed worldwide disarmament, with implementation vested in the federations. Thus in the armaments area, research and development focuses not on producing more and more potent devices for killing and destruction, but on finding effective and reliable ways to verify that such devices are not produced by any state or enterprise in the world.

There has been a major de-escalation in the civic use of weapons. Thanks to improved social conditions and more equitable functioning in the world's economy, criminality and violence are generally at a low level. There is less resentment and hate. The overall inaccessibility of handguns and other weapons has reduced the incidence of gang wars, massive killings, robberies, and rapes, and organized and maverick forms of criminality. Thus there is no need for large, highly equipped police forces and high-security prisons. With the exception of special forces, law-enforcement officers are equipped much as twentieth century English policemen were—with rubber sticks and handcuffs, occasionally supplemented by temporarily disabling noninvasive lasers.

The most important advances in technological development took place in the field of energy production. In view of the unsolved problem of nuclear waste disposal, we have eliminated massive reliance on nuclear reactors, although they are in use in major urban and industrial complexes. Our use of fossil fuels has been drastically reduced. Most of our energies are generated locally, by private dwellings and small communities. They come from the sun, as a direct source through photovoltaic and solar thermal technologies in private habitations and public buildings. Larger plants make use of hydropower, wind, wave, tidal, and geothermal energy.

Our entry into the "solar age" not only has opened up a practically infinite source of energy without polluting air, land, or water, but also helps re-balance the world's economy, since solar radiation reaches everywhere, and is particularly intense in the tropical and subtropical regions where the less industrialized economies are concentrated.

We are deeply concerned with the right choice of agricultural technologies. We refrain from massive mechanized solutions, preferring smaller-scale, locally based systems capable of maintaining biological diversity and producing healthy foods without the use of chemicals. We realize that the human body is part of nature and natural foods are best suited to maintain its health and vigor.

We do not raise millions of cows and pigs for the sole purpose of slaughtering them to provide meat for our table. We view such practices as morally reprehensible and irresponsibly wasteful of water, energy, and of productive land that could grow food for people instead of feed for herds of animals.

In addition to food production, our agriculture is a source of energy and industrial raw materials. We grow plants for producing fuels without encroaching on the production of humanly needed staple foods. Hemp and other plants are grown as a renewable raw material for the production of paper, textiles, and oil, as well as of some biodegradable varieties of plastics.

A key objective is the availability of a sufficient supply of clean water for all states and communities. Improved technologies of desalinization convert water from the seas into drinking water, drawing energy from sun, wind, and tide.

Our transport technologies aim at reconciling the requirement for mobility with the requirement for personal safety and public health. This is much less of a problem than it was at the turn of the century, for the emphasis on local self-reliance and autonomy reduced the need for people and goods to move long distances. In this regard the valuation accorded to the integrity of nature has been an important factor: it made us aware that energy, even if renewable, is to be used with care, and that transport systems have an unavoidable negative impact on the ecology. This impact is limited, but not entirely eliminated, by the use of clean renewable energies, such as liquid hydrogen, electricity, plant fuel, fuel cells, compressed air, and various hybrid motive technologies.

Our technologies of communication are advanced, but they are not substantially different from those of the early years of the century. Hardware is smaller, cheaper, and more powerful, and software is both simpler and more effective, adapted to use by people in all walks of life. Computers are at work in many facets of daily life and work. They eliminate some chores and make others easier, but they don't revolutionize existence in the way technological forecasters and science-fiction writers had envisaged. We still live on Earth in human communities and the embrace of nature. We make use of our technologies to live better and live more sustainably.

Advances in health-related technologies make a further contribution to the quality of life. Invasive medical procedures are limited to cases of birth defects, accidents, and serious malady. A softer and more holistic approach predominates in most other cases. The accent is on "salutogenesis": safeguarding health through the prevention of disease rather than using artificial means to cure diseases that are already acute. This requires that we view the human being as an integrated whole of body and mind, an integral part of society, culture, and environment.

The techniques and technologies that foster inner growth are an offshoot of our holistic approach to health. These "soft technologies" combine traditional practices such as meditation with biofeedback, EEG-analysis, and other state-of-the-art biomedical and psychophysical methods. They are recognized instruments of personal growth and development, and are widely accepted and used.

THE WAY WE GROW

A Joint Concluding Comment by the Community Counselor and the Schoolteacher

If you ask us what is the most important achievement of the world of 2032, we reply that it is to have learned to live and grow peacefully and sustainably. Growth is a basic attribute of life, a meaningful aspiration for humans, the same as for other species. But on a small planet not every form of growth is positive and morally permissible. Indiscriminate growth leads to conflict and can catalyze confrontation. We have learned to grow in a better way.

Extensive Growth

For most of recorded history, humankind has been growing the wrong way. The consequences became evident in the late twentieth century and turned dramatic in the beginning of the twenty-first. Earlier generations grew horizontally along the surface of the planet, growing by conquest and colonization. This was *extensive* growth. The main goal was to extend dominion over larger and larger areas of the planet.

Traditionally, the means to achieve extensive growth was conquest: the conquest of nature and the conquest of other, weaker or less power- and domination-oriented peoples. Successful conquest led to the colonization of tribes, nations, cities, and empires, who were subjugated to the ambitions and interests of the conqueror. For centuries this growth was accomplished by force of arms. Starting in the second half of the twentieth century it was also attempted by economic means: wealthy states and global companies used their power to impose their will and values on ever wider layers of the population.

For states the goal of extensive growth was territorial sovereignty, including sovereignty over the human and natural resources of the territories. For aggressive and unethical states the goal included extending their territory by force of arms: annexing the territory of neighboring states. The parallel goal for business companies was to extend their reach over more and more people, states, and resources.

We can subsume the ends of extensive growth under three "Cs": *conquest, colonization,* and *consumption.* Extensive growth concentrated wealth and power in fewer and fewer hands. It destabilized and destroyed established social structures and traditional cultures, and impaired vital ecological balances.

Intensive Growth

There is a better way to grow. Instead of extensive growth, we can opt for *intensive* growth. The ends of intensive growth can also be grasped by three "Cs," but three different "Cs": *connection, communication,* and *consciousness.*

One of the great myths of the industrial age was the separation of people from each other. This was legitimized by the worldview of classical physics. Like the mass points of Newton, humans appeared to be self-contained, mutually independent chunks of organic matter only superficially related to each other and to the environment. Physics no longer supports this view. We know that every particle is connected with every other particle throughout the universe, and every organism is connected within the web of life throughout the biosphere.

The connections that link people with each other and with nature can grow and evolve. One of the goals of intensive growth is to order and facilitate this process, creating syntony and coherence in place of random proliferation.

The second and third goals of intensive growth are directly linked with the first. They are to deepen the level of our connections by fostering communication, and the continued evolution of our consciousness. The full potential of human communication unfolds when the communicators comprehend the strands of connection among them.

Science has shown that we are linked with each other and with nature through the unified field, a cosmic energy and information field that recalls the ancient notion of Akasha. We learned that it takes an evolved consciousness to access this field and experience the subtle but real strands of communication that link us to each other and to the cosmos. The evolution and spread of an "Akashic" consciousness has helped us to grow without creating conflict between peoples and cultures, and doing harm to the environment.

INSIGHTS FOR TODAY

It's time to draw our conclusions. What have we learned from our visionary thought experiment?

The most important insight we have gained is that if we launch a WorldShift by the end of 2012, new values and a new consciousness could emerge by the year 2032. Two decades of living in a postshift era could enable a new generation of motivated people to evolve cooperative values and a transpersonal consciousness, creating lasting foundations for peace and sustainability in the world.

A shift in values and consciousness changes the way people manage themselves and the environment. A society where the members perceive their connections with each other and with nature has vastly improved chances of survival and development. It's more inclusive, embracing people from different ethnic groups, races, and religions in the search for the common good. It's more participatory, enabling all segments of the population to have a voice in making the decisions that affect our life. And it's more anticipatory, assessing the merit of today's decisions in light of their effect on people and the environment, not only next year and the year after that, but in the span of several generations.

Creating a sustainable world is feasible—in principle. Creating it in practice depends on what we do today: whether we evolve our consciousness, and begin to change consciously and in time.

SEVEN

Coming Home

A CREDO

The self-centered values and materialistic consciousness that dominated the world for the past two centuries have led us to a state of global emergency. But these values and consciousness are not a permanent feature of human nature: they are an aberration. For most of the twenty or thirty thousand years that our species has possessed a higher form of culture and consciousness, humans didn't define their interests in purely material terms, and they didn't think of themselves as separate from the world that surrounded them. They lived in the conviction that there is more to the world than matter and possessions made of matter, and that the world is one and humans are an intrinsic part of it.

The radical separation of a thinking, feeling human from an unthinking and unfeeling world came only with the modern age, and came mainly in the West. It prompted the uninhibited exploitation of nature by the thinking and feeling, and therefore superior, human species.

People with deep insight have never accepted this narrowly anthropocentric view, whether they were artists, poets, mystics, or scientists. Giordano Bruno, Leonardo da Vinci, Galileo Galilei, Isaac Newton, Nicolas Copernicus, and in more recent times Albert Einstein, have given eloquent testimony to their belief that the world around us, though it remains in many respects mysterious, is intrinsically and integrally whole.

Scientists have now discovered that the universe is indeed an inter-connected whole. The kind of space- and time-transcending connection that occurs in the sub-microscopic world of quanta is also found in the everyday world of the living, and even in the cosmos at large. It renders organisms, ecologies, the whole biosphere, and the universe itself, instantly and multidimensionally coherent.

Scientists have also discovered that our connections to each other have bona fide physical basis: care for others is not contrary to human nature, it's an integral part of it. This is the implication of the actual discovery of "mirror neurons" in our brain (previously their existence had been only theorized on the basis of the behavior of primates and from data furnished by functional magnetic resonance imaging). Mirror neurons fire not only when we are doing or experiencing something, but also when somebody around us does. Seeing another person grieving or joyful triggers the same response in our brain as our own grieving and joyfulness. Compassion and empathy are not just something we delude ourselves with, and at best are capable of—they are woven into the fabric of our nervous system. This makes perfect sense: if we weren't hard-wired for connecting with the feelings and experiences of others, genuine community life could never have evolved—we would still be living in self-centered bands, competing for resources and survival.

The insight that dawns is that all things are subtly but effectively tuned to all other things, and in some respects act as one. This has been known for thousands of years in the world's religious and spiritual traditions. For religions and spirituality the key words have been *love* and *oneness,* and for scientists they are now *connection* and *coherence.* In the final count, they mean the same thing. There is connection ("love") among all the particles, atoms, and galaxies in the universe, and the coherence ("oneness") it produces underlies all processes of evolution. There is no evolution in this universe but co-evolution, and co-evolution can only happen when the evolving partners are in syntony—when they are connected and coherent with each other.

With the new discoveries—or rediscoveries—in the sciences, we reach a different concept of who we are. We are not skin-enclosed, material parts in a vast machinelike universe, but interconnected organic elements of a living cosmos. In us nature's penchant toward structure, order, and coherence has reached a pinnacle unparalleled in this corner of the galaxy. Only in the last few hundred years has modern civilization suppressed our natural bent for empathy and coherence in favor of the self-centered pursuit of material interests. We may not be the only or even the most evolved form of life in the cosmos but, notwithstanding our faults and follies, we are one of the remarkably evolved ones. Our highly evolved brain and consciousness are eyes through which the cosmos can come to know itself.

Possessing the human kind of consciousness is a unique privilege, and it confers a unique responsibility. It allows us to rediscover ourselves in the cosmic order of things, and it gives us the moral obligation to do so. When we perceive our place in the universe we come to know our role and our mission: to be truly one with the world of which we are an intrinsic part.

The eye of the cosmos that has opened through us cannot be allowed to close. It must remain open and gain in depth and clarity. We can and must live up to the potentials of a species that possesses a high level of consciousness: to create love and connection, coherence and oneness, in and with the world. We can, and now we must, come home to this Earth—and the universe.

WorldShift 2012—A New Beginning

José Argüelles

The program is set. The future is already beginning. This is due to the magnitude of the event, WorldShift 2012. Its ripples are creating the first waves, washing up on the shores of the old consciousness. What is behind December 21, 2012, that it would create such an effect?

According to the time science of the ancient Maya, history is shaped by the influence of a galactic beam that Earth has been transiting for more than 5,100 years. A great moment of transformation awaits us in 2012, at the beam's end. The Mayan time science, popularly known as the Mayan calendar, is a mathematical code based on a science of holonomic resonance rather than atomic physics. The primary intention of the Mayan calendar system was not to measure time but to record the harmonic calibrations of the 5,125-year synchronization beam.*

December 21, 2012 marks the conclusion of the passage of our solar system through this galactic beam, which constitutes the *wave harmonic of history* for humanity. Why? The beam, 5,125 years in

---

*5125 years is equal to 5200 *tun* (a tun is a cycle of 360 days).

diameter, commenced August, 13, 3113 B.C.E., a date marked by the Mayan calendar long count as 13.0.0.0.0, which corresponds to 4 Ahau (a reckoning based on twenty day weeks, in which each day is named, and Ahau is the first day). This precise date—13.0.0.0.0, 4 Ahau—will occur again on December 21, 2012. The interval from 3113 B.C.E. to 2012 C.E. comprises the totality of history as we know it—from the First Dynasty of Egypt to the Twin Towers—hence the *wave harmonic of history*. During this cycle, humanity has gone from a tribal creature just learning how to live in cities to being a full-blown planetary organism.

The conclusion of the cycle in 2012 C.E. (kin1, 872,000, 13.0.0.0.0) bodes nothing less than a major evolutionary upgrading of the planetary life process. At this point, a resonant frequency phase shift will occur, presaging the brilliance of the post-2012 era of our galactic-solar-planetary realization: "Leaving the galactic synchronization beam in 2012 C.E., the cycle complete will be the cycle begun, and it will be as if we have seen ourselves for the first time and at the same time we shall recognize ourselves as human no more."[*] That is, we shall understand that we have passed into not only a post-historic but also a post-human, or superhuman, phase of our evolution.

What is this galactic beam and what is it synchronizing? It is an intelligently focused, high frequency time beam that is calibrated by thirteen sub-cycles called *baktuns*. Each sub-cycle is approximately 394.5 solar years, or 144,000 days, in duration. Thus 3113 B.C.E. to 2012 C.E. equals 1,872,000 days (13 baktuns or 144,000 days each). Each baktun is further divided into 20 smaller cycles called *katuns*, 260 in all. Each baktun was holographically charged with a program for activating and synchronizing the collective human DNA and its mental capacity into slow but steady increments of expansion and acceleration—a process characterized by the three "C's": conquest, colonization, and consumption.

---

[*]José Argüelles, *The Mayan Factor* (Rochester, Vt.: Bear and Company, 1987), 129.

During the twelfth baktun, 1224 to 1618 C.E., all of the major civilizations reached an apex of expansion and pre-industrial complexification, spilling over from the Old World to the New with the European conquest of the Americas and circumnavigation of the Earth. This set the stage for the thirteenth and final baktun, 1618 to 2012 C.E. During this concluding baktun, the cumulative effects of the first twelve cycles have attained an exponential momentum, known as the "climax of matter."

Not only did the year 1618 inaugurate what is usually referred to as the "scientific revolution," but it also marked the beginning of the mechanization of time and consciousness. This fact more than any other sets the final baktun apart from all of the previous ones. For the mechanization of time, through the perfection of the mechanical clock, creates alienation from nature, and a highly advanced capacity for social complexification and technological acceleration unlike anything previously known.

The process of the mechanization of time also created a totally unconscious mental-perceptual field in which the human being systematically separated itself from nature for the purpose of creating a vast industrialized order eventually to be known as the technosphere—a sphere or bubble of artificial time cast over the biosphere. The human species has been living on its own artificial time, apart from the rest of the biosphere that continues to operate in the natural cycles.

Artificial time is a double-edged sword. On the one hand, immersion in artificial time allowed the human species to construct a fantastically elaborate and complicated global civilization. On the other hand, separation from the natural cycles of universal order fostered the creation of a profoundly materialistic belief system. The resulting pursuit of resources and profits has crippled the biosphere, while fomenting a psychology of alienation that has resulted in the mega crisis of history so brilliantly described in part 1 of *WorldShift 2012*.

In the thirteen-baktun analysis, the human race is viewed as a single planetary organism. In entering the artificial time of the thirteenth baktun, human DNA became an excited and agitated field, extruding technology much like a spider extrudes its web. The purpose of technology and the pursuit of materialism is to allow the previously dispersed human community to come together, however chaotically, to finally realize itself as a single life-form capable of existing anywhere in the biosphere. This end result is the cumulative effect of the synchronization beam.

The process of arriving at this globalized condition has occurred so rapidly that the human mind—with its various provincial values shaped by antagonistic tribal, religious, and nationalistic beliefs—has hardly had a chance to rise up from the conflict it has engendered to see that we are in actuality one organism. Despite the present day crisis, "Little do the humans realize how close they are to the moment when the genetic game board of their reality becomes the illumined design of galactic destiny."*

There are other factors to consider in reviewing the significance and meaning of 2012. One relates to the perception of time. According to the Mayan view, time is the universal factor of synchronization. As the frequency shift occurs, the old calendar will be replaced by a new one based on the harmonic standard of thirteen moons of twenty-eight days. The universe will be revealed as an ever-evolving harmony, which will give rise to the new value of *time is art* instead of *time is money*. This fundamental shift will engender a new collective human priority: instead of ransacking the Earth for resources, we shall seek to transform it into a work of art. This perception will be of inestimable value in shifting our post-2012 priorities.

Importantly, the intensity of interest in the 2012 date is itself a manifestation of the process of acceleration and synchronization engendered by the galactic beam. Until the *Mayan Factor* was published in 1987, when the 2012 date was first dropped into the mass consciousness, next to no one knew about it. The purpose of the *Mayan Factor* was to alert people to the conclusion of the cycle of history in 2012, and the

---

*Argüelles, *Mayan Factor,* 154.

tremendous shift in consciousness this date augured. In the ensuing years, curiosity about the date developed, but gradually.

However, since 2007 interest in 2012 has become a feature of the mass consciousness, inclusive of books, websites, documentaries, and feature-length Hollywood films—at least two of them due to be released within the coming year. To many people it is the end of time, the end of the Mayan calendar, even the apocalypse.

These are popular misperceptions that unfortunately get raised to the status of supernatural reality by the entertainment industry. But the mass interest, whether fearful or hopeful, is already a shift in consciousness. Something is going to happen. But what happens is up to us. That is why *WorldShift 2012* is so significant as a handbook for this great change: it is the kind of positive co-creative script the world needs in order to conclude and regenerate the cycle on a successful evolutionary note.

The galactic synchronization foreseen as the conclusion of the cycle in 2012 is the moment of an evolutionary shift, or mutation. In fact the entire 5,125-year cycle—but an instant of geological time—could be seen as a mutative phase. The mutative phase complete, a new evolutionary stage begins. This new stage is known as the *noosphere,* a phase of evolution dominated by consciousness. In fact, what we refer to as the crisis are but the sundry effects of the *biosphere-noosphere transition,* the chaotic and dissipative shift into the new order of planetary reality. As a planetary organism, we are now being inevitably pushed into a new condition of *planetary consciousness.*

The Club of Budapest's "Call for a Planetary Consciousness" states that "Unless people's spirit and consciousness evolves to the planetary dimension, the processes that stress the globalized society/nature/system will intensify and create a shock wave that could jeopardize the entire transition toward a peaceful and cooperative global society. . . . Evolving human spirit and consciousness is the first vital cause shared by the whole human family. . . . Each of us must start with ourselves to evolve our consciousness."

This process, already underway, is an aspect of our evolutionary mutation. As a critical mass develops, it will snowball into a consciousness shift, the most primary prerequisite for entering the noosphere and creating the peaceable world envisioned in "Breakthrough 2032" in part 3.

Finally, in consideration that Earth is a part of the heliosphere, 2012 augurs a new solar age, the coming sixth sun of consciousness. Exploring the relation between solar frequencies and our own brain waves, which the Mayans term *tinkinantah*, the more adventurous members of our race will evolve a science of bio-solar telepathy, establishing an assured means for our continuing evolution. By creating a planetary telepathic network—based on an evolved Akashic consciousness—the notion of the noosphere as the mental sheath of the planet will be fully realized. Akashic consciousness, described by Laszlo as "a consciousness that recognizes our connections to each other and to the cosmos . . . a consciousness of connectedness and memory . . . [that] conveys a sense of belonging, ultimately, of oneness . . . a wellspring of empathy with nature and solidarity among people"—will be the supreme legacy of WorldShift 2012.

APPENDIX 1

WorldShift
Recommended Reading

Compiled by David Woolfson

State of the World

IPCC. *Climate Change 2007: Fourth Assessment Report.* Cambridge: Cambridge University Press, 2007.

UNEP/GRID-Arendal. *Planet in Peril: Atlas of Current Threats to People and the Environment.* United Nations Publications, 2006. See www.grida.no/publications/planet%2Din%2Dperil.

Worldwatch Institute, ed. *State of the World 2008: Toward a Sustainable Global Economy.* New York: W. W. Norton & Company, 2008.

Worldwatch Institute, ed. *Vital Signs 2007–2008: The Trends That Are Shaping Our Future.* New York: W. W. Norton & Company, 2007.

World Sustainability

Doppelt, Bob. *Leading Change Toward Sustainability.* Pensacola, Fla.: Greenleaf Publications, 2003.

Doppelt, Bob. *Power of Sustainable Thinking.* London: Earthscan Publishers, 2008.

Hawken, Paul. *The Ecology of Commerce.* New York: HarperCollins, 1994.

Henderson, Hazel. *Ethical Markets: Growing the Green Economy.* White River Junction, Vt.: Chelsea Green, 2007.

Heinberg, Richard. *Peak Everything*. Gabriola Island, B.C.: New Society Publishers, 2007.

McKibben, Bill. *Deep Economy: The Wealth of Communities and the Durable Future*. New York: Owl, 2008.

Sachs, Jeffrey. *Common Wealth*. New York: Penguin Press, 2008.

Senge, Peter, et al. *The Necessary Revolution*. New York: Doubleday Business, 2008.

Stern, Nicholas. *The Economics of Climate Change*. Cambridge: Cambridge University Press, 2007.

World Peace and Security

Ferencz, Benjamin. *Planethood*. Coos Bay, Ore.: Love Line Books, 1991.

Human Security Center. *Human Security Report 2005*. New York: Oxford University Press, USA, 2005.

Jacobs, Didier. *Global Democracy*. Nashville, Tenn.: Vanderbilt University Press, 2007.

Kay, Sean. *Global Security in the Twenty-first Century*. Plymouth, Devon: Rowman & Littlefield Publishers, 2006.

Ribbelink, Olivier, ed. *Beyond the U.N. Charter*. The Hague, Asser Press, 2008.

Roche, Douglas. *Global Conscience*. Toronto: Novalis Press, 2008.

Schell, Jonathan. *The Seventh Decade: The New Shape of Nuclear Danger*. New York: Holt Paperbacks, 2008.

———. *The Unconquerable World*. New York: Holt Paperbacks, 2004.

World Futures

Diamond, Jared. *Collapse*. New York: Penguin, 2005.

Glenn, Jerome and Theodore Gordon. *2008 State of the Future*. Washington D.C.: Millennium Project, 2008.

Homer-Dixon, Thomas. *The Upside of Down*. Washington D.C.: Island Press, 2008.

IPCC. *Climate Change 2007—Impacts, Adaptation, and Vulnerability*. Cambridge: Cambridge University Press, 2008.

Laszlo, Ervin. *The Chaos Point*. Charlottesville, Va.: Hampton Roads Publishing, 2006.

———. *Quantum Shift and the Global Brain*. Rochester, Vt.: Inner Traditions, 2008.

Mack, Timothy, ed. *Creating Global Strategies for Humanity's Future*. Bethesda, Md.: World Future Society, 2006.

Martin, James. *The Meaning of the 21ˢᵗ Century*. New York: Riverhead, 2007.

McKenna, Terence. *The Invisible Landscape: Mind, Hallucinogens, and the I Ching* (with Dennis McKenna). New York: Seabury, 1975.

Pearce, Fred. *When the Rivers Run Dry: Journeys Into the Heart of the World's Water Crisis*. Toronto: Key Porter Books, 2006.

———. *With Speed and Violence: Why Scientists Fear Tipping Points in Climate Change*. Boston: Beacon Press, 2008.

Petersen, John. *A Vision for 2012: Planning for Extraordinary Change*. Golden, Colo.: Fulcrum Publishing, 2008.

Smil, Vaclav. *Global Catastrophes and Trends: The Next Fifty Years*. Cambridge: MIT Press, 2008.

Walker, Brian, and David Salt. *Resilience Thinking*. Washington D.C.: Island Press, 2006.

Wright, Ronald. *A Short History of Progress*. Toronto: House of Anansi Press, 2004.

World Transformation

Anthony, Marcus. *Integrated Intelligence*. Boston: Sense Publishers, 2008.

Argüelles, José. *The Mayan Factor*. Rochester, Vt.: Bear & Co., 1996.

———. *Time and the Technosphere*. Rochester, Vt.: Inner Traditions, 2002.

Bornstein, David. *How to Change the World*. New York: Oxford University Press, USA, 2007.

Braden, Greg, et al. *The Mystery of 2012*. Boulder, Colo.: Sounds True, 2007.

Brown, Lester. *Plan B 3.0: Mobilizing to Save Civilization*. New York: W. W. Norton, 2008.

Chopra, Deepak. *Peace Is the Way*. New York: Harmony, 2005.

Elgin, Duane. *Promise Ahead*. New York: William Morrow & Co., 2000.

Goerner, Sally, Robert Dyck, and Dorothy Lagerroos. *The New Science of Sustainability*. Chapel Hill, N.C.: Triangle Center for Complex Systems, 2008.

Harman, Willis. *Global Mind Change*. San Francisco: Berrett-Koehler, 1998.

Hawken, Paul. *Blessed Unrest: How the Largest Movement in the World Came into Being and Why No One Saw It Coming*. New York: Viking USA, 2007.

Korten, David. *The Great Turning*. San Francisco: Berrett-Koehler, 2006.

Laszlo, Ervin, and Jude Currivan. *CosMos: a Co-Creators Guide to the Whole-World*. London: Hay House, 2008.

Mandell, Faye. *Self-Powerment: Towards a New Way of Living*. New York: Dutton, 2003.

Monbiot, George. *The Age of Consent*. New York: Harper Collins, 2004.

Petit, Patrick, ed. *Earthrise: the Dawning of a New Civilization in the 21ˢᵗ Century*. Munich: Herbert Utz Verlag GmbH, 2008.

Ray, Paul, and Sherry Anderson. *The Cultural Creatives: How 50 Million People Are Changing the World*. New York: Three Rivers Press, 2001.

Russell, Peter. *Waking Up in Time*. Updated edition. New York: Origin Press, 2009.

Senge, Peter, Otto Scharmer, Joseph Jaworski, and Betty Flowers. *Presence: Human Purpose and the Field of the Future*. New York: Doubleday, 2005.

Speth, James Gustave. *The Bridge at the Edge of the World*. New Haven: Yale University Press, 2008.

Westley, Frances, Brenda Zimmerman, and Michael Patton. *Getting to Maybe: How the World Is Changed*. Toronto: Vintage Canada, 2007.

Yunus, Muhammad. *Creating a World Without Poverty: Social Business and the Future of Capitalism*. New York: PublicAffairs, 2008.

The Club of Budapest Mission

THE MANIFESTO ON PLANETARY CONSCIOUSNESS

Drafted by Ervin Laszlo and the Dalai Lama and adopted at a meeting at the Hungarian Academy of Sciences in Budapest on October 26, 1996

THE NEW REQUIREMENTS OF THOUGHT AND ACTION

1. In the closing years of the twentieth century, we have reached a crucial juncture in our history. We are on the threshold of a new stage of social, spiritual, and cultural evolution, a stage that is as different from the stage of the earlier decades of this century as the grasslands were from the caves, and settled villages from life in nomadic tribes. We are evolving out of the nationally based industrial societies that were created at the dawn of the first industrial revolution, and heading toward an interconnected, information-based social, economic, and cultural system that straddles the globe. The path of this evolution is not smooth: it is filled with shocks and surprises. This century has witnessed several major shock waves, and others may come our way before long. The way we shall cope with present and future shocks will decide our future, and the future of our children and grandchildren.

2. The challenge we now face is the challenge of choosing our destiny. Our generation, of all the thousands of generations before us, is called upon to decide the fate of life on this planet. The processes we have initiated within our lifetimes and the lifetimes of our parents and grandparents cannot continue in the lifetimes of our children and grandchildren. Whatever we do will either create the framework for reaching a peaceful and cooperative global society and thus continuing the grand adventure of life, spirit, and consciousness on Earth, or set the stage for the termination of humanity's tenure on this planet.

3. The patterns of action in today's world are not encouraging. Millions of people are without work; millions are exploited by poor wages; millions are forced into helplessness and poverty. The gap between rich and poor nations, and between rich and poor people within nations, is great and still growing. Though the world community is relieved of the specter of superpower confrontation and is threatened by ecological collapse, the world's governments still spend a thousand billion dollars a year on arms and the military and only a tiny fraction of this sum on maintaining a livable environment.

4. The militarization problem, the developmental problem, the ecological problem, the population problem, and the many problems of energy and raw materials will not be overcome merely by reducing the number of already useless nuclear warheads, nor by signing politically softened treaties on world trade, global warming, biological diversity, and sustainable development. More is required today than piecemeal action and short-term problem solving. We need to perceive the problems in their complex totality, and grasp them not only with our reason and intellect, but with all the faculties of our insight and empathy. Beyond the powers of the rational mind, the remarkable faculties of the human spirit embrace the power of love, of compassion, and of solidarity. We must not fail to call upon their remarkable powers when confronting the task of initiating the embracing, multifaceted approaches that alone could enable us to reach the next stage in the

evolution of our sophisticated but unstable and vulnerable sociotech-
nological communities.

5. If we maintain obsolete values and beliefs, a fragmented consciousness, and a self-centered spirit, we also maintain outdated goals and
behaviors. And such behaviors by a large number of people will block
the entire transition to an interdependent yet peaceful and cooperative
global society. There is now both a moral and a practical obligation
for each of us to look beyond the surface of events, beyond the plots
and polemics of practical policies, the sensationalistic headlines of the
mass media, and the fads and fashions of changing lifestyles and styles
of work, an obligation to feel the ground swell underneath the events
and perceive the direction they are taking: to evolve the spirit and the
consciousness that could enable us to perceive the problems and the
opportunities—and to act on them.

A CALL FOR CREATIVITY AND DIVERSITY

6. A new way of thinking has become the necessary condition for responsible living and acting. Evolving it means fostering creativity in all people,
in all parts of the world. Creativity is not a genetic but a cultural endowment of human beings. Culture and society change fast, while genes
change slowly: no more than one half of one percent of the human genetic
endowment is likely to alter in an entire century. Hence most of our genes
date from the Stone Age or before; they could help us to live in the jungles
of nature but not in the jungles of civilization. Today's economic, social,
and technological environment is our own creation, and only the creativity of our mind—our culture, spirit, and consciousness—could enable us
to cope with it. Genuine creativity does not remain paralyzed when faced
with unusual and unexpected problems; it confronts them openly, without prejudice. Cultivating it is a precondition of finding our way toward
a globally interconnected society in which individuals, enterprises, states,
and the whole family of peoples and nations could live together peacefully, cooperatively, and with mutual benefit.

7. Sustained diversity is another requirement of our age. Diversity is basic to all things in nature and in art: a symphony cannot be made of one tone or even played by one instrument; a painting must have many shapes and perhaps many colors; a garden is more beautiful if it contains flowers and plants of many different kinds. A multicellular organism cannot survive if it is reduced to one kind of cell; even sponges evolve cells with specialized functions. And more complex organisms have cells and organs of a great variety, with a great variety of mutually complementary and exquisitely coordinated functions. Cultural and spiritual diversity in the human world is just as essential as diversity in nature and in art. A human community must have members that are different from one another not only in age and sex, but also in personality, color, and creed. Only then could its members perform the tasks that each does best, and complement each other so that the whole formed by them could grow and evolve. The evolving global society would have great diversity, were it not for the unwanted and undesirable uniformity introduced through the domination of a handful of cultures and societies. Just as the diversity of nature is threatened by cultivating only one or a few varieties of crops and husbanding only a handful of species of animals, so the diversity of today's world is endangered by the domination of one, or at the most a few, varieties of cultures and civilizations.

8. The world of the twenty-first century will be viable only if it maintains essential elements of the diversity that has always hallmarked cultures, creeds, and economic, social, and political orders as well as ways of life. Sustaining diversity does not mean isolating peoples and cultures from one another: it calls for international and intercultural contact and communication with due respect for each other's differences, beliefs, lifestyles, and ambitions. Sustaining diversity also does not mean preserving inequality, for equality does not reside in uniformity, but in the recognition of the equal value and dignity of all peoples and cultures. Creating a diverse yet equitable and intercommunicating world calls for more than just paying lip service to equality and just tolerating each other's differences. Letting others be what they want as long as they

stay in their corner of the world, and letting them do what they want "as long as they don't do it in my backyard" are well meaning but inadequate attitudes. As the diverse organs in a body, diverse peoples and cultures need to work together to maintain the whole system in which they are a part, a system that is the human community in its planetary abode. Different nations and cultures must now develop the compassion and the solidarity that could enable all of us to go beyond the stance of passive tolerance, to actively work with and complement each other.

A CALL FOR RESPONSIBILITY

9. In the course of the twentieth century, people in many parts of the world have become conscious of their rights as well as of many persistent violations of them. This development is important, but in itself it is not enough. In the 21st century we must also become conscious of the factor without which neither rights nor other values can be effectively safeguarded: our individual and collective responsibilities. We are not likely to grow into a peaceful and cooperative human family unless we become responsible social, economic, political, and cultural actors.

10. We human beings need more than food, water, and shelter; more even than remunerated work, self-esteem, and social acceptance. We also need something to live for: an ideal to achieve, a responsibility to accept. Since we are aware of the consequences of our actions, we can and must accept responsibility for them. Such responsibility goes deeper than many of us may think. In today's world all people, no matter where they live and what they do, have become responsible for their actions as:

- private individuals
- citizens of a country
- collaborators in business and the economy
- members of the human community
- persons endowed with mind and consciousness

As individuals, we are responsible for seeking our interests in harmony with, and not at the expense of, the interests and well-being of others; responsible for condemning and averting any form of killing and brutality; responsible for not bringing more children into the world than we truly need and can support; and responsible for respecting the right to life, development, and equal status and dignity of all the children, women, and men who inhabit Earth.

As citizens of our country, we are responsible for demanding that our leaders beat swords into ploughshares and relate to other nations peacefully and in a spirit of cooperation; that they recognize the legitimate aspirations of all communities in the human family; and that they do not abuse sovereign powers to manipulate people and the environment for shortsighted and selfish ends.

As collaborators in business and actors in the economy, we are responsible for ensuring that corporate objectives do not center uniquely on profit and growth, but include a concern that products and services respond to human needs and demands without harming people and impairing nature, do not serve destructive ends and unscrupulous designs, and respect the rights of all entrepreneurs and enterprises who compete fairly in the global marketplace.

As members of the human community, it is our responsibility to adopt a culture of non-violence, solidarity, and economic, political, and social equality, to promote mutual understanding and respect among people and nations whether they are like us or different, and to demand that all people everywhere should be empowered to respond to the challenges that face them with the material as well as spiritual resources that are required for this unprecedented task.

And as persons endowed with mind and consciousness, our responsibility is to encourage comprehension and appreciation for the excellence of the human spirit in all its manifestations, and for inspiring awe and wonder for a cosmos that brought forth life and consciousness and holds out the possibility of its continued evolution toward ever higher levels of insight, understanding, love, and compassion.

A CALL FOR PLANETARY CONSCIOUSNESS

11. In most parts of the world, the real potential of human beings is sadly underdeveloped. The way children are raised depresses their faculties for learning and creativity; the way young people experience the struggle for material survival results in frustration and resentment. In adults this leads to a variety of compensatory, addictive, and compulsive behaviors. The result is the persistence of social and political oppression, economic warfare, cultural intolerance, crime, and disregard for the environment. Eliminating social and economic ills and frustrations calls for considerable socioeconomic development, and that is not possible without better education, information, and communication. These, however, are blocked by the absence of socioeconomic development, so that a vicious cycle is produced: underdevelopment creates frustration, and frustration, giving rise to defective behaviors, blocks development. This cycle must be broken at its point of greatest flexibility, and that is the development of the spirit and consciousness of human beings. Achieving this objective does not preempt the need for socioeconomic development with all its financial and technical resources, but calls for a parallel mission in the spiritual field. Unless people's spirit and consciousness evolves to the planetary dimension, the processes that stress the globalized society/nature system will intensify and create a shock wave that could jeopardize the entire transition toward a peaceful and cooperative global society. This would be a setback for humanity and a danger for everyone. Evolving human spirit and consciousness is the first vital cause shared by the whole of the human family.

12. In our world static stability is an illusion; the only permanence is in sustainable change and transformation. There is a constant need to guide the evolution of our societies so as to avoid breakdowns and progress toward a world where all people can live in peace, freedom, and dignity. Such guidance does not come from teachers and schools, not even from political and business leaders, though their commitment and roles are important. Essentially and crucially, it comes from each

person himself and herself. An individual endowed with planetary consciousness recognizes his or her role in the evolutionary process and acts responsibly in light of this perception. Each of us must start with ourselves to evolve our consciousness to this planetary dimension; only then can we become responsible and effective agents of our society's change and transformation. Planetary consciousness is the knowing as well as the feeling of the vital interdependence and essential oneness of humankind, and the conscious adoption of the ethics and the ethos that this entails. Its evolution is the basic imperative of human survival on this planet.

State of Global Emergency
THE WORLDSHIFT 2012 DECLARATION

THE CRISIS AND THE OPPORTUNITY

There is no doubt that we are now in a state of global emergency. This unprecedented worldwide crisis is a symptom of a much deeper problem: the current state of our consciousness; how we think about ourselves and our world. We have the urgent need, and now the opportunity, for a complete rethink: to reconsider our values and priorities, to understand our interconnectedness, and to shift in a new direction, living in harmony with nature and each other.

Every person, community, and society in the world is already, or will soon be, affected by the global crisis, through climate change, economic breakdown, ecosystem collapse, population pressure, food and water shortages, resource depletion, and nuclear and other threats. If we continue on our present unsustainable path, by mid-century the Earth could become largely uninhabitable for human and countless other life-forms. However, total-system collapse caused by ecocatastrophes or escalating wars triggered by religious, geopolitical, or resource conflicts could occur much sooner.

These threats are real. The underlying causes of the present global crisis have been building momentum for decades and could soon become irreversible. Estimates of when the point of no return will be reached have been reduced from the end of the century, to mid-century, to the next twenty years, and recently to the next five to ten years.

The window of opportunity for pulling out of the current crisis and breaking through to a peaceful and sustainable world may be no more than a few years from now. This timeline coincides with the many forecasts and prophecies that speak of the end of the current cycle of human life on this planet and the possible dawning of a new consciousness by the end of the year 2012.

Today millions of forward-thinking groups and individuals all over the world are addressing the many opportunities presented at this critical time. Designs for sustainable systems, structures, and technologies are being developed and implemented in all sectors, at all levels, and in every society. This global awakening is a hopeful sign of the vitality of the human spirit and our ability to respond to the dangers we now face with insight and creativity.

The totality of our current efforts does not yet match the scope, scale, and urgency of the necessary transformation. If we now collaborate and act with vision, foresight, and commitment we can lay the foundation for a global community that is both peaceful and sustainable. We may then ensure our survival and well-being as well as that of future generations. While the window of time is still open, our top priority as global citizens is to accelerate our evolutionary shift to a planetary consciousness and, together, build a peaceful, just, and sustainable world.

AN URGENT CALL

We accordingly issue this urgent call to all the peoples of the world to deepen our awareness of both the dangers and the opportunities of the global crisis. We declare our firm commitment to work together to bring about a timely and positive WorldShift for the survival and well-being of all the peoples of the human community and the flourishing of all life on Earth.

The Club of Budapest Initiatives

For more information on the initiatives and projects of the Club of Budapest, see www.clubofbudapest.org.

THE WORLDSHIFT NETWORK

On the 12th of June 2007, Ervin Laszlo's 75th birthday, the Club of Budapest WorldShift Network was established as an international foundation. The purpose of the foundation is to connect the many organizations and individuals throughout the world who are presently working for a values-based holistic civilization and to strengthen their effectiveness.

The WorldShift Network (www.worldshift2012.org) focuses on six topics taken as organic components of an ongoing change of consciousness:

1. Unity of Humanity and Nature (shifting from domination to unity)
2. Global Subsistence Economy (shifting from exploitation to sustainability)
3. Salutogenesis (shifting from disease control to the safeguarding of health)
4. Global Wisdom Culture (shifting from ego-centered knowledge to holistic wisdom)

5. Participative Civil Society (shifting from domination to participation)
6. Planetary Peace and Freedom (shifting from separation to community)

THE WORLDSHIFT ALLIANCE

The WorldShift Alliance is to serve as a "mega-network" of like-minded individuals and organizations actively addressing our shared societal and ecological challenges and opportunities. The Alliance's broad mission is: *To effectively address humanity's growing challenges and opportunities with new thinking and actions in sufficient time to adapt to rapidly changing world conditions.*

Through a multifaceted "web community" the WorldShift Alliance is to bring the various sectors of society and the diverse cultures of the human family together with the thousands of individuals and groups who are actively working for a better world. It is to function as a worldwide community of global citizens and forward-thinking organizations who can together shift the human future to a peaceful and sustainable path.

THE WORLD WISDOM COUNCIL
In Partnership with the World Commission on Global Consciousness and Spirituality

The World Wisdom Council is to represent the collective wisdom of humanity, both masculine and feminine, and from every continent, major culture, and religion. Its core mission is to transcend narrow national agendas and self-serving individual interests, recognizing that thinking based on these levels cannot meet today's growing global challenges. It has been convened in the conviction that the paramount requirement in this age of discontinuity and transformation is to recognize that, through the development of a new dimension of consciousness, the world can be constructively changed by women and men wherever they live and whatever their interests and lot in life.

The WWC is an independent body, with initial members drawn from the Club of Budapest and the World Commission on Global Consciousness and Spirituality.

THE GLOBAL PEACE MEDITATION/PRAYER DAYS
In Partnership with the World Peace Prayer Society

Numerous tests and experiments have shown that deep prayer and meditation can heal people, heal other species, and create peace and harmony in human communities. The annual Global Peace Meditation/ Prayer Days were created to amplify the power of meditation to reduce the level of conflict and violence in the world and help create deeper understanding, tolerance, and readiness to live in peace with our neighbors near and far, as well as with nature.

The First Global Day, Sunday, May 20, 2007, brought together an estimated one million meditators in sixty-five countries on five continents. Never before have so many people in so many countries and from so many faiths and cultures joined together to direct the power of their meditation and prayer to the first truly common cause of all of humanity: peace on Earth.

The Mission of the Alliance for a New Humanity

Founded by Deepak Chopra

Our mission is to connect people, who, through personal and social transformation, aim to build a just, peaceful, and sustainable world, reflecting the unity of all humanity. We are people from all regions of the world and all walks of life who are joined by a common vision; to strengthen and sustain an actively compassionate humanity. Our movement is open to anyone who believes that creating a better world is possible—a world where all beings are valued equally, where Earth is revered and protected, and where the awesome potential of humanity can unite to bring about true peace and harmony.

Our intention is to create an alliance of people based on the awareness of humanity's interconnectedness. We believe that if enough people share the value of peace, war can be brought to an end. If enough people shift their awareness toward social justice, human rights, and environmental sustainability, then injustice, oppression, and the destruction of the ecosystem can be stopped. Such a shift is already occurring—now it needs critical mass, which in turn needs organization. The Alliance for a New Humanity aims to connect individuals, caring communities, and groups at a global level.

About the Author

Ervin Laszlo is founder and president of The Club of Budapest, founded in 1993, with a membership of many of the leading thinkers, artists, and visionaries in the world (see pages 98–114 for more on the Club of Budapest). He is founder of the think tanks the World Wisdom Council and the General Evolution Research Group. He is president of the WorldShift Network, a fellow of the World Academy of Arts and Sciences, a member of the International Academy of Philosophy of Science, and a senator of the International Medici Academy. He is the editor of the international periodical *World Futures: The Journal of General Evolution*.

Ervin Laszlo holds a Ph.D. from the Sorbonne and is the recipient of four honorary Ph.D.s. At the age of nine, he made his public debut as a child prodigy on the piano and by age fifteen was performing throughout the world. He received the Peace Prize of Japan (the Goi Award) in 2002, the International Mandir of Peace Prize in Assisi in 2005, and was nominated for the Nobel Peace Prize in 2004 and 2005.

A former professor of philosophy, systems science, and futures studies in the United States, Europe, and the Far East, he is the author, coauthor, or editor of eighty-five books translated into as many as twenty-two languages, including a four-volume *World Encyclopedia of Peace*. He lectures worldwide and lives in a converted four-hundred-year-old farmhouse in Tuscany.

Index

BOOKS OF RELATED INTEREST

Science and the Akashic Field
An Integral Theory of Everything
by Ervin Laszlo

Quantum Shift in the Global Brain
How the New Scientific Reality Is Changing Us and Our World
by Ervin Laszlo

The Akashic Experience
Science and the Cosmic Memory Field
by Ervin Laszlo

Science and the Reenchantment of the Cosmos
The Rise of the Integral Vision of Reality
by Ervin Laszlo

Morphic Resonance
The Nature of Formative Causation
by Rupert Sheldrake

Original Instructions
Indigenous Teachings for a Sustainable Future
Edited by Melissa K. Nelson

Transcending the Speed of Light
Consciousness, Quantum Physics, and the Fifth Dimension
by Marc Seifer, Ph.D.

Science, Soul, and the Spirit of Nature
Leading Thinkers on the Restoration of Man and Creation
by Irene van Lippe-Biesterfeld with Jessica van Tijn

Inner Traditions • Bear & Company
P.O. Box 388
Rochester, VT 05767
1-800-246-8648
www.InnerTraditions.com

Or contact your local bookseller